Weather Headliners

by

Rick Vaughan

Dorrance Publishing Co
585 Alpha Drive
Suite 103
Pittsburgh, PA 15238
Visit our website at www. dorrancebookstore.com

ISBN: 979-8-89211-191-1
eISBN: 979-8-89211-689-3

Introduction to Weather Headliners

This book is about meteorology, the branch of earth sciences whose primary goal is to understand, explain and forecast weather events. Meteorology is one of the five basic areas of earth science study, the other four being geology, climatology, astronomy and oceanography. Meteorology is the only earth science that most people actually talk about. Everyone knows about the weather. Ask someone about today's weather forecast, and chances are they will know all about it. Throw a little snow, freezing rain or a strong cold front into the forecast, and people's interest increases exponentially. I wrote Weather Headliners for people like that, those who are fascinated by the weather and want to learn and understand more. And why would anyone want to learn more about meteorology? Simply because there is so much to learn. It has been said thar there is more left to be understood about the weather than is understood today . And there is so much variety in the weather; so much change hour-to-hour, day-to-day, week-to-week. In this book you will learn about rain, snow and wind and many other features of the weather. There are different chapters on subjects such as hurricanes, lightning the water cycle, heat waves and floods. What is special about the book is that many topics one rarely sees in print like the wind rose, how snow melts, the use of paleontology for helping define the earth's temperature, and Thomas Jefferson's incredible weather diary are discussed.

Weather Headliners is a collection of my experiences with the weather, a personal accumulation of what I have observed, read about, been exposed to and otherwise learned throughout many decades of interest in meteorology. It was really only when I retired that I could devote a large percentage of my time to the weather. One of the things I did with that extra time was to write Weather Headliners. For years I wrote a limited issue monthly article on some aspects of the weather, everything from Nor'easters to lake effect snow to how we measure the earth's temperature. Weather Headliners is easy to read, and full of both interesting tidbits and unexpected surprises. So, for instance, an article on how raindrops are formed is followed by articles on the history of roofs and umbrellas. Similarly, the intricacies of snowflake crystallization are contrasted with articles on what makes snow melt, avalanche and snow fences. Weather Headliners reads both like a textbook and a novel.

Meteorology has been a part of my life since I was just a kid. I made a rudimentary weather station from household items like dixie cups and rubber bands when I was just eleven, and followed that up with a plaster-of-paris facsimile of the Water Cycle. I took advantage of the weather experiences that accompanied the especially active weather of the 1950's and 1960's in New England. That weather featured unusually heavy snow, cold temperatures and hurricane landfalls and aftermaths. In 1987 I set up a sophisticated amateur weather station that allows me to see current weather parameters and create tables and graphs. It even has a neat self-emptying rain gauge.

Weather Headliners has a unique style and a more relaxed presentation Hopefully you will find yourself smiling at some of the commentary and eager to turn the page. Weather Headliners is not a textbook. Basically, if I didn't understand it, or couldn't explain it, it's not in the book. Weather Headliners consists of eight standalone Chapters, each representing a weather feature (like Rain) or a process (such as Forecasting). The Chapters are Rain, Snow, Wind, Storms, Climate, Global Warming, Forecasting and Risk. Each Chapter is independent, associated with but not directly connected to any other Chapters. The chapters can be read in any order without loss of understanding, though the current order is appealing. This is also true of the order of the subheadings in each Chapter.

In the postscripts at the end of the book there are several items that may be of interest. The first is a complete listing of all eight Chapters and the subheadings associated with each.

There is a somewhat romantic essay called "A Young Boy's Fascination with the Weather", a look back at the author's boyhood in Milford, Connecticut. There are two segments that aptly apply to the author's sense of the beauty of the Weather, one called "Characteristics of the Weather" and secondly a slightly

offbeat interpretation of various weather phenomena based on the author's observations over six decades of observing the weather, and called "What I Know".

The Postscript also serves to document a "List of References that were used
in the creation of the book, and some folks whose efforts I want to acknowledge
for their contributions to the creation of Weather Headliners

I hope you enjoy reading the book as much as I did writing it.

Rick Vaughan

Acknowledgments – Weather Headliners

Thanks to *Steve Zubrick* for strengthening the technical content of *Weather Headliners*. Steve's tireless efforts ran all the way from suggesting a few words to improve technical understanding, to complete rewrites to bring the book's technical content up-to-date and in line with current meteorological thinking. Steve earned a Master of Atmospheric and Earth Sciences from Old Dominion University, Norfolk, VA in 1986. After several years in the private sector, Steve joined the National Weather Service (NWS) in 1986. Just five years later Steve was named one of the first Science and Operations Officers (SOO), this at the NWS local forecasting offices in Sterling, VA , serving Northern Virginia and Dulles Airport. The SOO was created to ensure that all field office personnel were up-to-speed with the latest information regarding climate, forecasting, weather and associated systems. Steve is a past president of the National Weather Association, and participated in numerous governmental and research committees. Steve retired from the NWS in 2022 and is currently an adjunct professor in Meteorology at George Mason University in Fairfax, Virginia. Because of Steve's efforts, contributions and above all his knowledge of meteorology , *Weather Headliners* is a technically credible publication.

My wife *Mary* brought strategic thinking to the project, keeping us focused on options, some of which were initially discarded but ultimately made the presentation of higher interest and value to future readers. We purposely, on Mary's recommendation, minimized the use of bullets, underlining and pictures that often distract more than focus. Mary is credited with naming the book *Weather Headliners*, and with the "Chapter" concept for displaying and sharing meteorological information. She has also been a big help with the thankless task of proofreading hundreds of pages of text and revisions.

Table of Contents for Each of the Eight Chapters in Weather Headliners

CHAPTER 1 – RAIN

People need and use water for a seemingly endless number of activities that includes cooking, cleaning, irrigating crops and lawns, transportation and recreation. Rain provides the moisture that fills our rivers, reservoirs and aquifers. And because water is so prevalent and useful, mankind has always had a special relationship with Rain. Rain has been welcomed and detested. There have been times when we have wanted more rain and times when we have wanted less. Rain has been a threat and a godsend. It has provided comfort and pain. It has been revered and reviled. It is indeed hard to imagine life as we know it today without rain.

In this chapter on Rain, we will describe how cloud droplets and raindrops are formed, and what causes it to rain. We will review what happens when there is too much rain, and what happens when there is too little. We will walk through the long and interesting history of roof design and associated materials of construction improvements to keep themselves and their families dry, and sheltered somewhat from dangerous animals.

The study of cloud and raindrop formation at a scientific level is a complex and technically challenging undertaking. The principles of precipitation formation are among the key, core technologies of meteorology.

The prevailing thinking regarding the basic mechanics of raindrop formation is that first very tiny water droplets condense on even smaller dust, pollen, salt and smoke particles present in the atmosphere. These cloud particles, which may act as nuclei for future raindrop formation, are very small and very light. It might take 500 just to weigh an ounce. Accumulated together the water droplets are visible as clouds but are not heavy enough to fall to the earth as rain or another form of precipitation. The water cycle uses a relatively simple and straightforward model for many situations involving rain formation. It has proven effective and useful for describing and characterizing a number of applications ranging from monsoons to deserts. In its simplest form, the model has four components, which are

- Evaporation, which is the conversion of liquid to gas, in this case water to water vapor
- Lift, which is the movement of that water vapor up and into the troposphere propelled primarily by updrafts
- Condensation, which is the conversion of a gas to a liquid, in this case the creation of water droplets when the water vapor that is being "lifted" contacts the colder surfaces in the cloud, and
- Precipitation, which is any product of condensation that then falls toward and contacts earth.

It takes about <u>a million cloud droplets to come together to form just a single raindrop.</u> For raindrop creation to occur in sufficient and meaningful quantities that life will continue on Earth as we know it today, literally trillions of raindrops will have to be produced and dispersed within the clouds and then as raindrops. And this must take place relatively quickly. Existing processes like the simple "evaporate-lift-condense-precipitate" model described above cannot handle such a massive redistribution of moisture.

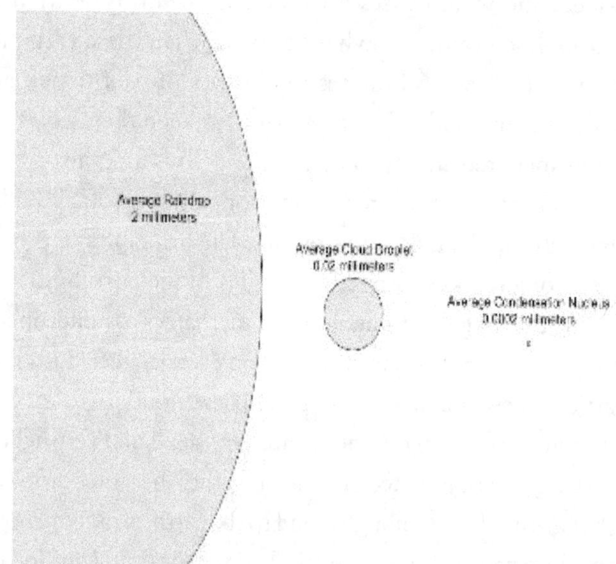

Scientists have, in general, agreed that there are two primary raindrop formation processes at work in the atmosphere are: 1) the Collision-Coalescence

Process CCP), and 2) the Ice Phase Process (IPP), the latter also the latter known as Wegener Bergeron Findeisen process.

The Ice Phase Process accounts for most global precipitation; the Collision Coalescence only where temperatures are above freezing. Outside of the tropics and warm seasonal locations, the vast majority of precipitation is created via the Ice Phase Process. ICC Phase Process theory is that snow and freezing rain are the precursors of raindrop formation. The frozen material encounters a warm zone upon descent and then is supercooled prior to impacting earth's surface. Collision-Coalescence is just as the name implies. Collisions between droplets causes growth when one droplet bumps into another. The larger of the two will "absorb the smaller" and grow larger still, continuing to grow until the raindrop gains sufficient weight to begin its descent to the ground. The only difference between a cloud droplet and a raindrop is that the raindrop has a greater discernable/non-negligible fall speed. And the larger the raindrop, the faster the fall. The above theories explain how this combination takes place. The fall velocity of small cloud droplets is determined by the Stokes Drag Law equation. A different equation is used for spherical raindrops to calculate raindrop terminal fall speed.

Amidst all the controlled chaos in the rain drop formation portion of the atmosphere, the raindrops continue to grow, generally by colliding with other raindrops and "borrowing" a little of their neighbors' condensate. As they fall toward the ground, the droplets are increasingly subject to resistance from the air through which they are dropping. Small raindrops have a diameter of less than 0.04". At this point most of the droplets generally round as they. As the drops become larger, the bottom of the raindrop is increasingly subjected to updrafts, which reduce the raindrops' descent speed and creates some interesting shapes. Everything from the top half of a hamburger bun to a parachute have been mentioned. As this is occurring many raindrops have become quite large and, at some non-specific point, begin their journey to earth's surface.

Several additional alternative processes have been proposed for the creation of individual raindrops. The Bergeron process is dependent on the growth of crystals. In due course a mixture of supercooled water droplets and larger ice particles fall into the lower, warmer parts of the atmosphere where they melt into raindrops. It has also been theorized that wind is an important factor governing rain drop formation, benefiting from the turbulent nature of the atmosphere. This turbulence creates winds that provides a much higher probability that collisions between cloud droplets will occur, accelerating the pace and extent of rain formation. The laws of physics keep the drops from becoming ridiculously large; the largest drops may split, forming new droplets. The raindrops themselves are exceedingly light in weight; A raindrop descending at 15 mph from a cloud base of 5,000 feet will reach the Earth's surface in about four minutes.

Air Pressure

There is an atmosphere on Earth and the weight of all the air above a point (say at the surface) is called "air pressure". At the Earth's surface, a hypothetical column of air exerts a force (or pressure) of about 14.7 pounds per square inch (psi). Pressure varies across the Earth because of uneven heating by the sun of various surfaces (land/water/and so forth). Air pressure depends on the temperature and density of molecules. Areas where the sun warms the air will have lower pressure because warm air is less dense (i.e., the molecules move around faster and occupy less space over a given volume) than colder air, and will tend to rise. Due to the Earth's rotation and associated Coriolis effect, winds in a low pressure area will tend to swirl in a counterclockwise manner in the northern hemisphere (opposite in the Southern Hemisphere). Thus, a low pressure system is where air is generally rising.

Rain Initiators

With rain being as omnipotent and omnipresent as it is, with rain occurring on all continents, and showing itself in so many different ways, everything from drizzle to torrential tropical downpours to sleet, one might think that there would be many ways to trigger rain formation. In fact there is a limited number, just three

- Low pressure systems and fronts
- Orographic precipitation
- Convection

Low Pressure Systems are formed when two air masses containing different temperatures and moisture levels meet. The air masses are separated by fronts. Colder, drier, denser air on one side of the front; warmer, more humid, lighter air on the other. The development and movement of low and high pressure systems and their accompanying fronts creates instability in the atmosphere.

The inherently lower pressure encourages warm air to rise and contact colder air, leading to condensation. Low pressure systems and their accompanying fronts are responsible for much of the world's rain. If the difference in temperature and humidity is dramatic, the low can strengthen, producing even more rain. If some rotation is added to the mix, severe weather may occur. Alternatively, high pressure stifles the rise of water vapor, generally leading to fair skies and little rainfall.

Orographic precipitation refers to rain produced primarily by geological features; more specifically changes in elevation such as occurs in mountainous terrain. The concept is that, given the right location, and wind speed and direction, that warm, humid air being propelled up the side of a mountain (often called an upslope wind) is subjected to cooling and condensation as the rising warm air meets colder air at elevation. Geological features don't change the processes for creating rain, but they do facilitate the process by accelerating the "lift" required for rain to occur.

Actually, there are two ways orographic precipitation occurs: 1) the forced ascent of moist air up an orographic feature like a mountain and 2) the differential daytime heating that causes air to flow up the mountain/hill slope. While most are probably familiar with the first method, e.g., the ascending moist flow off the Pacific Ocean into the US West Coast mountains, the second method on a more local scale is actually quite common over most mountains, especially during the warmest parts of the year. Orographic precipitation is a prevailing condition in the mountains of the U.S. West. Here a massive redistribution of water is underway as moisture-laden westerly winds blow onshore from the Pacific Ocean. Heavy rain blankets the windward sides of the mountains. By the time the winds are crossing over the ridge, most of the initial moisture has fallen. The orthographic process is very efficient. So, whereas the annual rainfall in Eugene, OR is 46", Bend, OR, on the other side of the mountains, gets just 11".

Convection is the most common of the processes that create rain. When the sun heats the earth, pockets of warmth are created. These air masses rise, cool and may eventually condense sufficiently to cause precipitation. Convective processes are quite vigorous, and can be intense with updraft speeds reaching perhaps 30 mph. However, showers can develop into more organized systems, like thunderstorms, which can then organize and grow "upscale" into even more organized precipitation systems like mesoscale convective systems, which is a collection of thunderstorms that act together as a system. Mesoscale events produce many of the more intense rainfall and flooding events.

So, precipitation will only occur when water vapor is lifted significantly high in the atmosphere to allow condensation to occur. Although all four ingredients of the rainfall model (evaporation to produce the water vapor that rises into colder areas of the atmosphere where condensation occurs, and precipitation is produced) must be "working" for rain to occur. The mere presence of the ingredients does not necessarily mean it will rain.

- All things equal, the more moisture the more rain, and vice versa.
- All things equal the faster the ascent and the more vigorous the upsloping winds, the more rain and vice versa.
- All things equal, the colder the atmosphere at higher altitudes the more rain, and vice versa.

The Water Cycle

In ancient times, man was convinced that the land masses were floating on the oceans and that rivers originated from under the earth. By the 4th century B.C., scholars began to question how rivers could continually flow into the oceans without the oceans overflowing. The concept of a rudimentary water cycle began to emerge. Writings increasingly reflected theories that tied together rivers and rain, clouds and oceans, and clouds and rain. Greek and Chinese scholars were the first to specifically describe a closed "water cycle" system. By the 15th and 16th centuries such theories had been accepted. The person credited with discovering the water cycle is Bernard Palissy, a French

Huguenot hydraulics engineer. In 1580, he stated, "Water evaporates from oceans, forms clouds which move inland and then condense and fall as rain." The world contains approximately 332,500,000 cubic miles of water. Almost 97% of this water is stored in the earth's oceans as salt water. The remainder of the world's water resides in rivers, lakes, ponds, streams, inland seas, and underground aquifers and in the earth's atmosphere. Despite the apparent enormity of the total volume of water on, and under, the earth's surface, these volumes pale in comparison to the non-water mass of the earth.

The Water Cycle describes the movement of water on, under and above the Earth's surface. It is also called the Hydrological Cycle.

- There are four main stages in the water cycle – evaporation, condensation, precipitation and collection (collection is sometimes referred to as accumulation). And there are other processes as well that one might say contribute to, perhaps round out, the cycle by making sure all the little contributors are recognized.
- For instance, some of the water in the water cycle moves primarily on or under the Earth's surface with no or little change in physical property, that is, for example, liquids stay liquids. Other primary processes include:
- Processes where there is no change in state
    - o Infiltration – the flow of water from the surface into the ground
    - o Subsurface Flow – the movement of groundwater within underground aquifers
    - o Percolation – the flow of water through soil and rock caused by gravity
- Some involve changes of state, like solid to liquid, and solid to vapor
    - o Snowmelt - water runoff caused by melting snow
    - o Sublimation, which is a change of state of snow or ice directly to vapor without passing through the liquid stage
- Water vapor lifted up and into the atmosphere through evaporative processes
    - o Transpiration – the release of water vapor from plants and soil into the air. A single large oak tree will transpire around 40,000 gallons of water a year

o Evaporation - transformation of water from liquid to gas as air passes over water

- Each year approximately 121,000 cubic miles of water enter the atmosphere from transpiration and evaporation (together known as evapotranspiration), 86% of it from the oceans

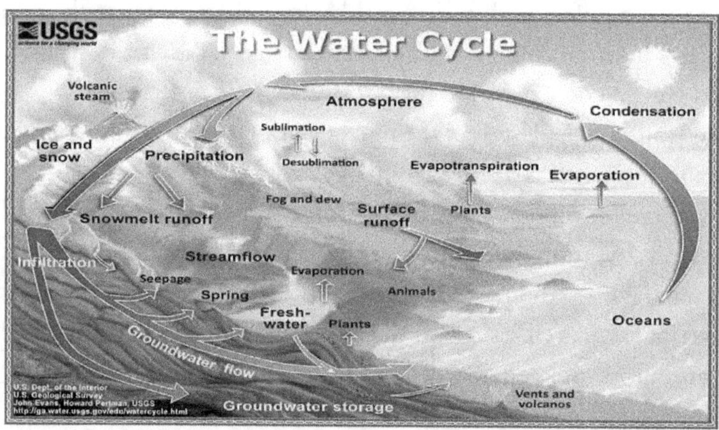

Sometimes lost amidst the fascination of the water cycle and its associated processes is one particularly crucial role that water plays during evaporation. When it comes to evaporation, water is totally ambivalent regarding where and what type of water is transformed to vapor. *Evaporation* handles all waters: salt, brackish, crystal clear water from Crater Lake, muddy water from pig farms. Only clean water vapor rises into the atmosphere. Almost all the salt is left behind, as are most of the dirt and other undesirable substances like volcanic ash and sand dust. Evaporation literally acts as nature's desalination plant. But, once evaporation has created water vapor and condensation has created precipitation, it is *rain* that is the most important *distributor* of water. Rain provides essential fresh water for the earth and all living things.

The water cycle is the journey water takes as it circulates from the land to the upper reaches of the atmosphere and back again. The sun's heat provides energy to evaporate water from the Earth's surface (oceans, lakes, rivers). Plants and dirt also lose water to the air (a process called transpiration). The water vapor eventually condenses when it meets cooler air, cloud formation and precipitation (rain, sleet and snow) are triggered, and water *returns to the land or sea. Some of the precipitation soaks into the ground. Some of the underground water, called groundwater, is trapped and <u>accumulates/collects</u> between rock or clay layers. Some of the water flows downhill as runoff (surface or subsurface), eventually returning to the oceans and seas.*

Types of Precipitation

Precipitation is <u>any form of water particle, liquid or solid</u>, that <u>falls from the atmosphere</u> and <u>reaches the ground</u>. Rain is by far the most common type of precipitation. Other types of precipitation include sleet, snow, freezing rain, hail, drizzle, and graupel.

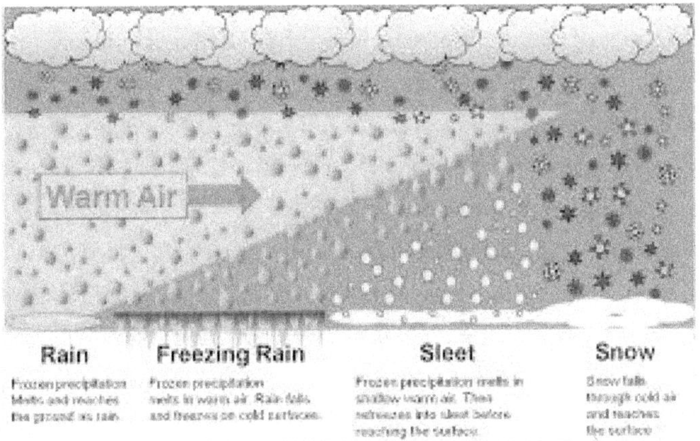

The model is of a warm front moving from left to right into an area of colder weather. A vertical slice of the atmosphere at any point along the horizontal can be used to ascertain the final state of that type of precipitation. The four major types of precipitation are:

<u>Rain</u> – precipitation stays in the warm sector from top to bottom – rain all the way down to earth

<u>Freezing Rain</u> – precipitation that may start as freezing rain encounters a sufficiently deep layer of above-freezing temperatures and completes its journey as a liquid. If temperatures at ground level are below freezing the falling rain becomes ice on all surfaces with which it comes in contact. In a worst-case scenario, the freezing rain continues for days, ice continues to build up

on tress and electrical transmission pylons and tremendous damage occurs when plants or structures can no longer withstand the weight of the ice. This scenario was played out during the 1998 North America Ice Storm that affected parts of Ontario and Quebec. Ice thicknesses of up to five inches led to catastrophic damages to homes, businesses and especially infrastructure.

Sleet (also called ice pellets) are produced when the air is not cold enough to support snowflake formation but cold enough for rain pellets to freeze on their journey to Earth.

Hail is almost exclusively associated with warm weather thunderstorms. It is produced by a fascinating and complex process.

Sleet and hail are meteorological terms that frequently cause confusion and are often incorrectly used. Sleet is simply frozen rain drops, most likely experienced during the colder/winter months. Hail is primarily a summertime/warm weather product of thunderstorms. Here is how hail is formed.

A small ice particle is free and heading for earth when its descent is interrupted by a strong updraft which pushes the ice particle upwards. This particle becomes coated with a layer of supercooled water which immediately freezes. The particle is heavier, but not strong enough to overcome the power of the wind. Up it goes again, lifted by strong air currents. Up/down/up/down. The cycle is repeated perhaps dozens of times. Finally, the hailstone gets heavy enough to break loose of the updrafts. The hailstones fall to the earth, where they may inflict serious damage to buildings, vehicles and crops.

Hail sized 1/4 inch or greater in diameter is typically associated with thunderstorm activity. Individual hailstones of an inch or more in diameter are indicative of very severe thunderstorms. Hailstone formation is clearly one of meteorology's more unique processes.

Drizzle is precipitation because, unlike fog and mist which don't, drizzle can actually create a small accumulation (that is, it meets the precipitation requirement of contacting the ground).

Graupel is a unique meteorological feature that occurs when supercooled water droplets adhere (i.e., freeze) onto existing ice crystal or snowflake. Graupel flakes are typically about 0.15" in diameter. Graupel are also called soft hail or snow pellets.

Graupel flakes are formed under conditions that don't occur frequently and most people have not seen them, So, Graupel snowstorms are very rare. A brief squall and some whitening of the ground are far more likely. Because the Graupel flakes are created under supercooled conditions, many of the small number of flakes that are formed do survive the trip to earth and can be observed even when the ambient temperature is above 40F.

Virga is an interesting meteorological phenomenon that does not qualify as precipitation. Meteorologically, virga is an observable streak of precipitation falling from a cloud that evaporates, or sublimates before reaching the ground. Virga occurs when the air has an extremely low dew point that encourages rapid evaporation. Virga oftentimes shows up on Radar as rainfall. Although it is in fact raining inside the clouds with which it is associated, that rain is actually not reaching the ground. So, Virga is not precipitation.

Measuring Rainfall

We collect rain data primarily to be able to better assess how our weather today compares to the climates of the past. In order to do this we must create, sustain and use accurate databases that allow us to confidently make those comparisons. Rainfall that occurred yesterday with rain that fell days ago or centuries ago. Data to see if Boston established a new record for 15-minutre rainfall intensity, or to refute claims that rainfall was increasing or decreasing because of global climate change. Or just to help people go about their everyday lives and routines. Do you need to have your irrigation system turned on today and programmed to water your lawn if you had an inch of rain just yesterday?

While collecting rainfall data is useful for climate studies, there are many, many uses for very short term (daily, less-than-a-day) rainfall data. For example, in a flash flood situation, the National Weather Service uses rainfall accumulation data over very shorter time spans (1/2/3/6 hours) to help forecasters assess flash flood potential. That data is needed in near real time.

The National Weather Service's Standard Rain Gauge (SRG)

The National Weather Service uses, as its standard, the 20" tall, 8" diameter Standard Rain Gauge (SRG). The gauge is a simple, non-recording instrument with four major components: measuring stick, overflow can, collector funnel and measuring tube. The 8" standard gauge used by all 122 Local Weather Forecasting Offices (LWFO) provides uniformity, continuity and credibility for weather data worldwide. Precipitation is measured in the measuring tube of the SRG with a black, laminated measuring stick approximately 24" long. The stick is graduated with easy to see marks. The overflow can come in handy when an accumulation exceeds 24" during the observation period. In addition to acting as an overflow rain collector, it is the location's primary collector of frozen precipitation. When snow, sleet or freezing rain are expected, the innards of the SRG are removed. The frozen precipitation collected by the overflow container is melted to 1) provide a liquid equivalency that ensures accuracy of monthly or annual "rainfall" and 2) to calculate the moisture level of the snow.

Flooding

For most of any given year, and for almost all years, there is a quiet equilibrium between Mother Nature and her lakes, streams and especially rivers. But if enough rain falls in a short enough period of time, flooding may well occur. Flooding can also be caused by snow/ice melt even with little rain contribution.

Close to 100 people die on average each year in the U.S. from all types of flooding combined. The chart below shows 2022's preliminary data for deaths from flooding, tornadoes and lightning. One observation is that fatalities from tornadoes and lightning have shown greater reduction in number of fatalities with time than has deaths from flooding.

	Fatalities 2022	Fatalities 10 yr Average	Fatalities 30 yr Average
Flooding	91	104	89
Tornadoes	71	45	71
Lightning	36	22	36

In the past 10 years (2013 to 2022), the average of flooding-related deaths has ticked up to 113. There is significant variability from year to year in fatalities, even in the 10-year average. The deaths ranged from a low of 50 in 2013 to 178 in 2016. To put things in some sort of perspective, flooding from Hurricane Katrina led to over 1500 deaths in 2005. Flooding also causes billions of dollars of damage to buildings, homes and infrastructure. River flooding and <u>flash flooding</u> are responsible for more damage to infrastructure, and <u>human misery</u> than lightning, <u>more than tornadoes, more than hurricanes</u>. A little less than half of all natural disasters <u>worldwide</u> involve floods. The power of moving water is very deceiving. Only 6" of fast moving water can knock a person off his or her feet; two feet of water will lift and carry a car.

There are three commonly accepted types of flooding: coastal flooding, flash flooding and river flooding.

<u>Coastal flooding</u> is often associated with storm surge, the wall of water that lashes the northeast quadrants of hurricanes and (and occasionally Nor'easters) as they come onshore. But coastal flooding can and does also occur without any storm surge. King tides (which is a popular, nom-scientific

term to describe exceptionally high tides), and persistent onshore flow are examples that commonly creates coastal flooding.

<u>Flash flooding</u>. As the name implies, flash flooding can occur in 15 minutes or less, often in response to summertime thunderstorms and can occur anywhere in the U.S. Humid summer air can rise quickly during hot summer days. If the air rises near a mountain range and is kept in place by winds in and around the storm, rain may fall non-stop for many hours. The water rushes quickly into the valleys below, many of which have surfaces baked as hard as concrete, funneling and accelerating the water, moving at speeds so high that people in its way hardly have time to react and can be swept away by the on-rushing water. Campers, who often select dry riverbeds as camp-sites, are particularly susceptible, but especially at night. <u>Flash flooding is the most destructive and deadly form of rain.</u>

A flash flood hit Shadyside, Ohio in June of 1990 when 4" of rain fell in 2 hours, creating a 30' wall of water which flowed through town, destroying virtually everything in its path and killing 26 people. Flash floods by their very nature are unpredictable, difficult if not impossible to accurately forecast, quicker than those trying to find and warn the vulnerable, and deadly. A localized heavy thunderstorm occurring miles away, often not within earshot, can create an incredibly fast flowing flash flood that strikes those in its way with amazing speed, power and destruction.

River Flooding

It doesn't take much to change the quiet equilibrium of a calm, meandering river to mayhem. River flooding is created by heavy rain falling steadily at frequently high rates over an extended period of time (days versus hours). Here are some memorable U.S. river floods.
- California January 1862
    - o Pummeled by over 20" of rain in just a month, the Sacramento River rose 24', exceeding all flood stages on record and spilling over its banks, creating a massive lake 250 miles long and 20 miles wide
- Ohio River March 1913

> o Four days of heavy rain, during which some locations like Belle-fontaine, OH received nearly a foot of rain, led to widespread flooding all along the Ohio River and its tributaries with Dayton, OH cresting 34' above flood level, and reaching a depth of 18 feet in some portions of town. The flood claimed 123 lives.

- Mississippi River April 1927
  > o One to two feet of rain, much of it occurring in a very short period of time, inundated a large area of the Mississippi River Valley from St Louis to New Orleans, an area just south of which recorded 20+" in just 18 hours. New Orleans was saved from destruction when President Herbert Hoover ordered the demolition of the Poydras levee. This allowed the Mississippi's flood waters to bypass the city and flow directly into the Gulf of Mexico.

Following these floods the Army Corps of Engineers was given responsibility for controlling the nation's major waterways.

<u>River flooding is the most common type of flooding.</u> River flooding occurs when more water enters an environment than can be absorbed into the soil or drained away to streams, rivers, lakes and other bodies of water. In most cases, events of this type can only be expected to occur in the presence of major storm systems that are stationary and/moving slowly.

The jet stream typically moves weather systems like fronts and high and low pressure systems across the country from west to east. During the summer months the jet that was such a force of change during the winter pulls back into Canada. There is virtually no movement. Stagnant conditions mean rain can continue to fall for very long periods of time on the same land areas until the storm finally gets "caught-up" in the jet stream and moves away. <u>Two major contributors to river flooding are the inland remains of tropical storms and the cut-off low pressure system.</u>

Cut-Off Lows and the Inland Phase of Tropical Systems

Unlike wind speed and intensity, rainfall from a tropical system does not diminish quickly once landfall is achieved. A number of atmospheric features help convert the rain system into a so-called extratropical cyclone, a low pressure system. A frontal boundary, a ripple in the jet stream, a blocking high pressure system, a secondary low pressure system. All of these meteorological features can create conditions that are optimal for heavy rain, in some cases many hundreds of miles from landfall. If conditions are favorable for condensation, prodigious rains can occur—sometimes 20" or more, in some cases over 30". No topography or man-made feature can deflect, absorb or channel such amounts of water. Serious flooding is inevitable. In just the past 50 years, literally dozens of landfalling hurricanes are actually remembered more for their rain and subsequent river flooding and accompanying deaths and damages to homes, businesses and infrastructure than they are for their winds, which were the initial focus of the storm and its cause for concern.

Here are some noteworthy hurricanes ultimately known more for their rain than their wind:

- Hazel 1954, Mid Atlantic
- Camille 1969, Central Virginia
- Floyd 1999, Mid Atlantic, New England
- Sandy 2012, Mid Atlantic, New England
- Camille 1969, Central Virginia
- Harvey 2017, Texas
- Florence 2018, North Carolina

This phenomenon of incredible inland, post landfall hurricane remnant rain is described superbly in the paper. *Second Wind: the Deadly and Destructive Inland Phase of East Coast Hurricane*s.

Little known, rarely discussed and a bit of a mystery, the Cut-Off Low can hold its own against any prolific rainmaker. Most weather in the continental U.S. is steered from west to east by the polar and/or sub-tropical jet streams. These conveyor belts of swiftly moving, high altitude air push, pull and nudge high and low pressure systems on their trips across the USA. So, if all is working per design, the weather, be it dry or wet or windy, doesn't last too long at any one spot on the map. When a low (or high) pressure system escapes from the grip of the jet stream, it is suddenly left quite alone, no longer being pushed or pulled or positioned. Because it is now completely separated from the Jet Stream, it is referred to as a Cut-Off Low.

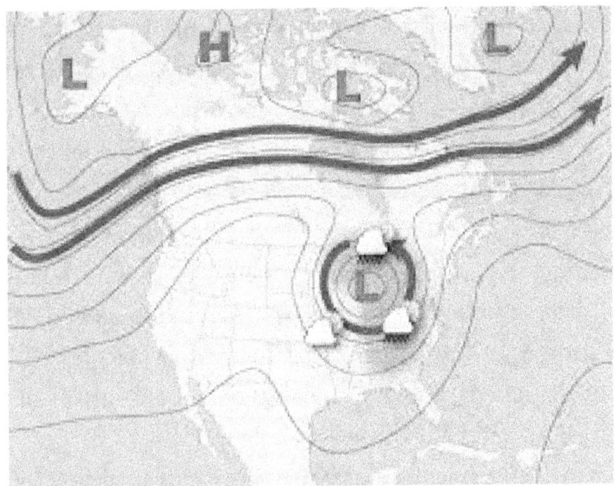

NOAA defines a Cut-Off Low as *"a closed upper-level Low which has become completely displaced (cut off) from the basic westerly current and moves independent of that current"*. The reality is that, left alone, Cut-Off Lows often become nearly stationary for days. If the location of the Low and the accompanying meteorological conditions are right, the persistent counterclockwise rotation associated with the Cut-Off Low draws in moisture that can lead to many days of heavy rain.

Florence, the second wettest storm in U.S. history, dropped some 10 trillion gallons of rain over a large area of primarily North Carolina in September 2018. Elizabethtown, NC had almost three feet of rain, and dozens more reporting stations came in with 30 or so inches. The record rainfall easily beat the previous NC record of 24.06" at Southport, North Carolina. The flooding in 2018 was exacerbated by heavy rains that preceded Florence, rendering the ground fully saturated and unable to accommodate any additional water. The highest rain totals occurred in the Cape Fear river basin and flood plain, and true to its designation as such, flood waters rose quickly, inundating thousands of roads and homes. Harvey (2017) hung around South Texas for days, delivering unprecedented quantities of rain. Three-day rainfall totals were set all across South Texas. Harvey's waters flooded over 300,000 structures and 500,000 cars, led to 68 fatalities, and caused over $125 billion damages. An average of 17.5" of rain fell on an area of 14,000 square miles, nearly three times the size of the State of Connecticut.

<u>Dams</u> are used to catch and retain water from rain and melting ice and snow. The water can then be used in a variety of ways from producing electricity to sail boating. The dams themselves are constructed from concrete, steel or earth. Earthen dams are generally cheaper and easier to construct but are not nearly as strong and impenetrable as steel dams. Water puts pressure on the dam walls from the moment in is introduced. With time the earth may become prone to softening and "flow". Most earthen dams have limitations defining maximum allowable capacity. This can be converted to an allowable height at a specific location using a yardstick. This makes questions like "How much as the volume of the lake changed in the past 24 hours?" or "Is there sufficient flow to continue to run the hydro-electric plant?"

A <u>Dam Failure is typically a catastrophic event triggered by the sudden, rapid and uncontrolled release of impounded water</u>. One of the most notorious dam failures occurred in Johnstown, Pennsylvania on May 31, 1889, following a night of torrential rain. The South Fork Dam, which held back Lake Conemaugh, had not been well maintained, and the added stresses and strains caused by the added water from the heavy rain led to the total collapse of the dam. Twenty million tons of water rushed through the valleys and canyons above Johnstown at speeds of 40 mph and into the town, killing more than 2,200 people.

A flood plain is an area of land adjacent to a stream or river which stretches from the banks of its channel to the base of the enclosed valley walls, and which periodically experiences flooding during periods of high water caused by rain and/or snowmelt. Man learned over the ages to depend on annual flooding of specific rivers to deposit much needed minerals and rich deposits of silt, sand and soil for sustaining agriculture and indeed civilizations. The Nile River in Egypt was, without doubt, the single greatest factor in sustaining civilization in that part of the world. Ironically enough, especially in the U.S., millions of Americans live directly on, or very close to, flood plains where their livelihood and their very safety are threatened by the river. Their risks are addressed by local authorities and the U.S. Army Corps of Engineers. In Egypt on the other hand, few levees have been built to in any way interfere in the annual floods that deposit life-giving silt onto the surface of the flood plain, ensuring a good harvest. Others could probably benefit from remembering that a flood plain is a portion of land on which crops, and livestock abound, but only a few people live. Plans are continually devised and implemented to increase the height of levees to hold back rivers from flooding the adjacent flood plains. Nonetheless Mother Nature periodically gets her way and flooding occurs in places named for such occurrences—of course, the flood plains.

Despite the fact that deserts cover about 20% of the Earth's landmass and that there are many locations on the earth that struggle to receive enough rain to survive, Mother Nature is capable of creating condensation that can lead to prodigious amounts of precipitation. Here are some example of extreme rainfall rates and totals experienced in the USA.

Where	When	What
Unionville, MD	July 4, 1956	1.23" in 1 minute
Alamogordo, NM	Jun 5, 1060	2.03" in 5 minutes
Central WV	May 4, 1943	13.80" in 1 hour
Smethport, PA	July 17-18, 1942	34.30" in 1 hour
Waipa Gardens, Kauai, HI	July 25-26, 1979	43.00" in 24 hours
Kukaiau, Hawaii	February 27 – March 6, 1902	62.00" in 4 days
Kukaiau, Hawaii	February 27 – March 2, 1902	82.00" in 8 days
Kukui, Maui, HI	March 1942	107.00" in 1 month
Kukui, Maui, HI	1982	704.83" in 1 year

The term "monsoon" refers to the seasonal change in the direction of the prevailing winds of a region. Monsoons are especially common in countries bordering the Indian Ocean in the months between May and October. In countries prone to monsoonal changes, rainfall can be substantial, indeed periodically catastrophic. During especially vigorous monsoonal seasons, areas of India may receive over 400 inches of rain, more than is needed for agriculture and more than can be handled by the existing infrastructure. The result can be flooding, damage to crops and loss of human life.

Sites in Colombia reporting extraordinary rainfall are located just five degrees above the equator in an area of extremely warm temperatures, high humidity and generally calm winds. Its location in the Doldrums features very low atmospheric pressure which provides the lift to convert water vapor to rain. Other cities like Bergen, Norway have measurable rain and snow on over 325 days a year; the difference is that the total amount of annual rain is appreciably less than what could result in amounts of rain of 400+ inches a year. For these amazing rain totals to occur, *the complex Collision Coalescence Process and the simple "evaporate-lift-condense-precipitate" model* had to be working at peak capacity. Mount Waialeale on Hawaii's island of Kauai receives over 450" of rain on an equally incredible 350 rainy days. Notable at Waialeale are very steep cliffs that cause moisture-laden air to rise rapidly, resulting in very high rates of condensation.

Deserts

Although there are some differences of opinion regarding how little moisture should be considered the limit, it is highly likely that, to be defined as a <u>*desert*</u>, *a given piece of the Earth's surface must have all three of these:*

- Very little precipitation (less than ten inches of rain on an annual basis)
  - o A situation in which evaporation exceeds rainfall
- Limited presence of vegetation and animals
  - o Only living organisms that have adapted to the harsh climates (virtually no water, withering heat) can survive
- Little ground water if any
- Deserts are generally associated with high temperatures;
- Places like Death Valley regularly reach temperatures that average 115F in July and August
  - o Both Alaska and Antarctica have deserts
- Deserts are found on every continent and cover about 20% of the world's surface area
- Deserts are home to about one billion people, more than 10% of the world's inhabitants

- The driest areas of the U.S. are the Desert Southwest, the Great Basin of Nevada, and central Wyoming
- Although the word "desert" may conjure up thoughts of shifting sand dunes, in fact dunes cover only about 10% of earth's deserts
    - o Deserts exist in all geographical areas, including on mountains
- Desertification is the term used to define the expansion of deserts to new locations
    - o Deserts expand primarily because of changes in climate, including long-term drought
    - o But farmers using poor agricultural techniques are also culpable
- The number of acres of land qualifying as desert is increasing, in some locations at an alarming rate
    - o Man is powerless to stop the advance of deserts

In recent years, some of the most rapidly expanding deserts have been in Africa, China and Kazakhstan.

There are five distinctive types of desert—subtropical, coastal, rain shadow, interior and polar.

<u>The exceptional dryness of each type of desert results from a unique set of geological, geographical and/or meteorological factors that greatly limit the opportunity for any condensation to occur.</u>

- Subtropical deserts are found along the Tropic of Cancer and the Tropic of Capricorn, 15 to 30 degrees north and south of the Equator, respectively. Desert-like conditions occur when air descends to fill in the space created after heavy rains near the equator move away from the equator toward the tropics. These air masses descend and warm up. The descending air hinders cloud formation, shutting off the ascending air necessary for rain formation. The world's largest hot desert, the Sahara, is a subtropical desert in North Africa. The Sahara is almost twice the size of the continental United States
- Coastal deserts owe their existence to cold ocean currents. Air blowing off the ocean toward the shore is chilled by contact with the cold water, often enveloping the land with a layer of heavy fog. Although there is plenty of water vapor available, the atmospheric updrafts that are nec-

essary for condensation, coalescence and raindrop formation are not present. The Atacama Desert, on the Pacific cast of Chile, is a coastal desert; cooling is provided by the cold Humboldt ocean current. Arica, located in northernmost Chile, has one of the lowest annual average rainfalls of any city in the world; no rain fell from October 1903 through January 1918—14 years. Some areas of the Atacama Desert go decades without rain; one weather station has not recorded a drop of rain in over 100 years. These same meteorological mechanics are at play in Southern California with the cold California current stifling rain development.

- Rain shadow deserts form when moisture has been squeezed out of an air mass on one side of a mountain range, and it is likely little moisture is left when it reaches the other side of the mountain. When moisture-laden air contacts hilly or mountainous terrain, the air is lifted up into colder air, condensing the available moisture, leading in many cases to significant precipitation on the upsloping side of the terrain. The land and its occupants on the down sloping side of the hilly terrain are said "to be in the shadow of the rain". Death Valley, in southern California, and a smidgen of Nevada, is a rain shadow desert of the Sierra Nevada mountains. Similar rain shadow effects are responsible for the very dry conditions in eastern Oregon and eastern Washington State.

- Interior deserts are found in the heart of continents far from any of the meteorological features like low pressure systems, fronts, and moisture-laden winds which are prerequisite for rain. These deserts exist because the moisture simply can't reach them. The Gobi Desert in China and Mongolia is hundreds of miles from the Pacific Ocean; a good source of moisture but just too far away.

- Polar deserts exist because the moisture needed for precipitation is locked up in the ice of artic regions. Parts of the Arctic, Siberia and Antarctica are classified as desert. There are vast quantities of water on, or just underneath, the surfaces of these polar regions. However, almost all of the moisture is locked up in glaciers, tundra or ice sheets year round, unavailable for precipitation. Global warming may change all of this.

- All deserts are beset by meteorological and/or geographic features that provide insufficient moisture or atmospheric lift. These can include:

o *Downdrafts that warm the atmosphere and stifle precipitation*

o *Cold ocean currents that provide moisture but no means of lift*

o *Polar regions where moisture is locked in frozen surface and sub-surface reservoirs*

o *Regions far from any frontal movements or other sources of moisture*

o *The leeward sides of hilly terrain where the majority of available moisture is converted to rain on the windward side of the terrain*

- On the other hand the rainiest places on earth often benefit from both copious amounts of warm, humid air and geological features that are oriented in a manner that maximize condensation

- The processes that dry out the land, empty reservoirs and wilt vegetation are slow, determined, incremental and relentless. Flash floods occur in a matter of hours, river floods over weeks, maybe months. Droughts can last for years.

- As long as the factors that favor the "evaporate-lift-condense-precipitate" model are missing, or not functioning effectively, it is highly unlikely that significant rainfall will result.

A Personal Story from an 8th Grader

It's been a very hot summer. It's funny how we didn't have very many of those big thunderstorms with thunder and lightning and lots of rain, like we usually have every summer. A lot of hot dry days are perfect for playing baseball before it gets too hot, and heading off to the pool when it is. It's September now and the weather is still beautiful, but the lawns are brown, and we are being told not to wash our cars; I guess there's a concern about wasting water. The leaves turned brown and fell from many of the trees early this year. Someone said that's what the trees do to protect themselves from dry conditions. Now we are hearing that the local government authorities have asked us to voluntarily cut back on water usage; everything from shorter showers to fewer flushes, and absolutely no lawn watering or car washing.

There are big fines if you get caught. The watermelons were small this year and so were the pumpkins.

We're almost at the end of the year. There's only been about 30% of normal rainfall since the spring and someone said they heard that means we are in something called a *drought*. Whatever it is called it seemed to sneak up on us, slowly, without warning till suddenly, well, we're in one! The farmers had a horrible harvest. The soil is just too dry. The farmers' water sources, like the little ponds that were always full as far back as I can remember, well they have simply dried up. The winter didn't change things at all. My uncle used to call ones like this "Open Winters". No snow. Heck, we couldn't even find enough snow in the mountains to go skiing. Meteorologists are keeping track of the drought—they are often the first to recognize when one starts.

The last six months of last year had a rainfall deficit of nearly 15" and January and February rainfall and melted snow just added up to 1.84" versus a normal of 4.51". February is the eighth consecutive month of below average precipitation. People are starting to get nervous. Rivers, our primary sources of water for drinking, showers, cooking and other important needs, are noticeably low. Rain isn't charging underground aquifers or keeping above ground reservoirs full. In fact, these are shrinking at an alarming rate. The rainfall continues to be less than average. The very dirt that needs rain is so dry that no moisture is available from evaporation to aid in the rain-making process. It's now over a year since it basically stopped raining. At school we are assigned a term paper on the drought. The research we do leads us to topics we don't understand, words like polar jet stream, El Nino and arctic oscillation that are apparently contributing to the dryness. We learn that droughts are complicated meteorological features, and that they occur periodically as part of the climate cycle. We are also assured that droughts do come to an end and that balance of sorts is re-established. One thing is for sure—one big snowstorm or thunderstorm is not going to be enough to end this long period of below normal precipitation. We will need sustained rainfall over months to rid ourselves of this drought. But it will end and then, who knows, maybe we'll be worrying about floods in a year or so.

Drought

- Drought is a period of time when there is significantly less rainfall than normal.
- The National Weather Service defines drought as "a deficit in moisture that results in adverse impacts on people, animals and vegetation over a sizeable area".
- Drought is a normal part of the climate cycle and occurs periodically in all parts of the country with individual episodes often separated by decades.
  - o Areas west of the Mississippi River and in particular the Great Plains and the Southwest are more prone to drought than the East.
- Drought is a slow moving hazard, a long-term process, often lasting many years.
  - o It is difficult to define when drought starts and when it is over.
- Recognizing drought before it intensifies can reduce impacts and save money.
  - o How one recognizes the impending drought depends on how it affects them personally.
- Drought leads directly to serious secondary weather events like dust storms.
- Drought means different things to different people, regions of our country and the world.
- Ultimately drought is in the eyes of the beholder.
- It really depends on each person's perspective.
  - o To a farmer who has just cultivated land and planted seeds a month of drought conditions and little to no rain could spell disaster. During the growing season three to four weeks without rain would have a similar effect.
  - o To the meteorologist keeping records and studying long term trends, drought is a prolonged period (perhaps two months or more) without significant precipitation.
  - o To a hydrologist, drought constitutes insufficient water to keep streams flowing and lakes full.

o And to a water resources manager, drought is any period of time that adversely affects the quantity and quality of water provided to the general public.

The United States Drought Monitor was formed in 1999 to provide the country with accurate, timely drought information. Every Thursday a revised map of the United States is posted, showing where drought is occurring, and its severity. The Drought Monitor is a joint production of the National Drought Mitigation Center at the University of Nebraska-Lincoln, NOAA and the Department of Agriculture (USDA). A map uses five classifications: abnormally dry (D0), showing areas that may be going into or coming out of drought and four levels of drought: moderate (D1); severe (D2); extreme (D3); and exceptional (D4). USDA uses the drought monitor to trigger disaster declarations and eligibility for low interest loans. The IRS uses it for tax deferrals on forced livestock sales. Decision makers use it to trigger drought responses. By comparing maps on successive weeks, one can ascertain growth, stability and/or diminishment of drought at any spot or area on the map. The "Monitor" <u>does not forecast</u> drought, it confirms it based on how much precipitation did or didn't fall up to the Tuesday morning before the Thursday release, giving the assessment team two days to fill in the maps. See below the Drought Monitor for October 14, 2022. California was stuck in severe drought; unbeknownst to them that an "atmospheric river" was poised to bring extraordinary rain and snow to the Golden State in just a few months, effectively bringing the drought to an end (at least for a while). Drought was further diminished when another major precipitation event occurred in 2023 with a visit from the remnants of Hurricane Hilary, first tropical cyclone to make it to California in decades,

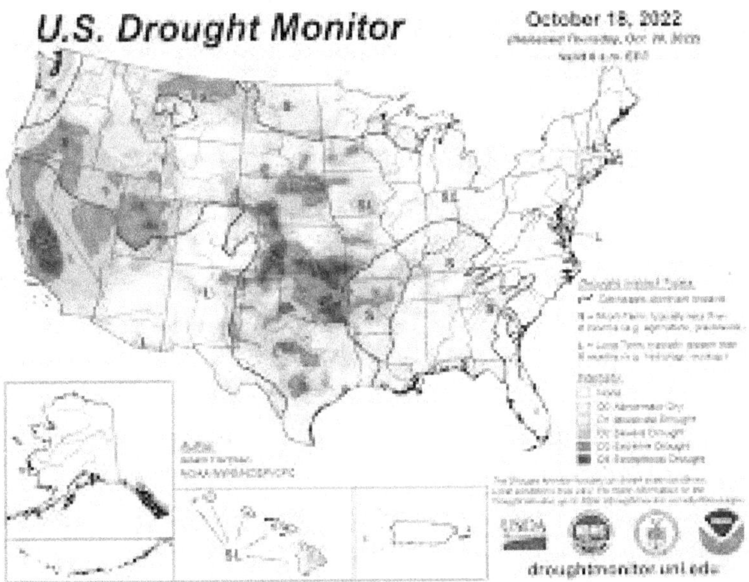

In addition to the Drought Monitor, the Palmer Drought Severity Index adds significantly to our understanding of drought severity. The Palmer Drought Severity process builds on a close relationship between itself and the Drought Monitor. Palmer develops its colorful "drought maps" by collecting and analyzing thousands of actual soil samples for moisture content. This provides the Monitor and Palmer with the ability to work collaboratively on drought situations, one using dry soil and the other one moist.

Heat Wave

Most parts of the contiguous 48 United States experience sustained hot weather in July and August. The few exceptions climatologically included Oregon and Washington. Over the past few years these states too have experienced hot summer weather. The higher elevations of the Rocky Mountains and, in the East, the Appalachians generally don't experience such hot spells. In almost all cases heat waves develop when the atmosphere becomes stagnant. High pressure systems strengthen, remain stationary for days or weeks, and become increasingly hotter. Along the East Coast of the United States, and in the absence of strong jet streams or other atmospheric steering currents, summertime high pressure systems tend to drift off the eastern seaboard and become stationary. These are the "Bermuda Highs", so named because of their proximity to the island. High atmospheric pressure inhibits the upward flow of moisture-laden air and suppresses the cloud formation that is the precursor of precipitation. The result is day after day of clear skies, and stagnant, hot conditions, with few meteorological features to provide cooling.

There are similarities between droughts and heat waves. Both create hot, dry, stagnant weather. The big difference between droughts and heat waves is time. Droughts develop slowly and can last for years, while heat waves are transitory, measured in weeks not months Heat waves can exacerbate droughts. As the high pressure strengthens, air sinks and traps the hot air at the surface, effectively shutting off normal radiative night-time cooling. The stronger and more statutory the High, the longer the air has to heat up during the day, the less time and driving force there is for cooling to take place at night, and the more severe is the heat wave.

Relief from heat waves can come from a wide variety of very welcome sources including an invigorated jet stream that nudges the high pressure along. The remnants of an East Coast tropical storm can carry much-needed rainfall northbound. A cooler air mass to the north with enough oomph can displace the hot air causing the heat wave. And, as summer wanes in late August, there are fewer and fewer minutes (and eventually hours) of time for the sun to heat the atmosphere.

No matter where you are on the earth's surface, a detailed definition for "Heat Wave" needs to first and foremost consider the temperatures normally experienced at that location. What is a heat wave for some is merely business as usual for others. Having said that, most locations consider three days a minimum requirement. Here are some normal climatological high temperatures for July for a few locations:

Washington, D.C.	89	Boston	82
Dallas	96	Las Vegas	104
Baghdad, Iraq	110	San Francisco	71
Half Moon Bay, CA	64	Copenhagen	69

At least one U.S. State and a number of countries use Heat Wave definitions to achieve this sought-after goal. Some of these are quite imaginative.

"Country"	*Days Above*	*This Temperature*
California ("Heat Storm")	Three or more	100F (Must encompass thousands of sq. miles)
The Netherlands	Five or more	77F
Belgium Luxembourg		(provided 3 of those days also exceeded 86F)
Denmark	Three or more	84F (when experienced by more than 50% of the country)
Sweden	Five or more	77F
Adelaide, Australia	Five or more, or three or more	95F 104F

Strictly speaking, the definition of a Heat Wave is based on 1) air temperature and 2) the length of time such abnormally high temperatures persist. But the story of heat waves is really about how hot you feel, not the ambient temperature. To determine how we feel, we have to include humidity (the amount of moisture in the air) when talking about heat waves. Humidity has a dramatic effect on human beings' discomfort. The human body relies on the evaporation of perspiration to keep itself cool. The evaporation process is in turn dependent on the amount of moisture in the air. If the air is saturated with moisture, evaporation cannot occur at a sufficiently rapid rate, and perspiration and associated discomfort persist. Dew Point, the temperature below which water vapor will condense into a liquid, effectively characterizes the discomfort associated with hot, humid weather. Unfortunately, Dew Point has never gained any traction as a weather parameter. Most folks not only don't know what it is, they can't relate any given dew point reading to their particular level of comfort or discomfort. To the rescue comes the Heat Index.

Very simply the Heat Index is what the atmosphere feels like to the human body when relative humidity is combined with the air temperature. The Heat Index is expressed as an air temperature, which most people can understand, relate to and compare with previous experiences. Generally the effect of humidity in the summertime is to create conditions that are less comfortable. That is, humidity creates a heat index higher than the temperature of the air alone. For example, when the air temperature is 92F, and the relative humidity is 55%, the resulting Heat Index is 109F. The Heat Index is also known as the Apparent Temperature or the Temperature-Humidity Index. Or you might see it referred to on the web, in a newspaper, or from your local TV weather personality in its most basic form—Dew Point.

NWS Heat Index   **Temperature (°F)**

Relative Humidity (%)	80	82	84	86	88	90	92	94	96	98	100	102	104	106	108	110
40	80	81	83	85	88	91	94	97	101	105	109	114	119	124	130	136
45	80	82	84	87	89	93	96	100	104	109	114	119	124	130	137	
50	81	83	85	88	91	95	99	103	108	113	118	124	131	137		
55	81	84	86	89	93	97	101	106	112	117	124	130	137			
60	82	84	88	91	95	100	105	110	116	123	129	137				
65	82	85	89	93	98	103	108	114	121	128	136					
70	83	86	90	95	100	105	112	119	126	134						
75	84	88	92	97	103	109	116	124	132							
80	84	89	94	100	106	113	121	129								
85	85	90	96	102	110	117	126	135								
90	86	91	98	105	113	122	131									
95	86	93	100	108	117	127										
100	87	95	103	112	121	132										

Likelihood of Heat Disorders with Prolonged Exposure or Strenuous Activity

☐ Caution   ☐ Extreme Caution   ☐ Danger   ☐ Extreme Danger

Man & Rain

For most of human history rain has been a mystery to mankind. Man did not understand how rain formed, what the water cycle was all about, or why there was plenty of rain during some growing seasons, but too much or too little during others. In this segment, we will look at how and what mankind developed to literally keep rain off their collective heads—the roof and the umbrella.

As with many natural phenomena, weather in general and rain in particular were handled by the gods and other mythological figures. In Greek mythology, Zeus was the god of weather, assisted by Poseidon, god of sea storms, and illuminated by Iris, goddess of the rainbow. Rome had Jupiter and Neptune for the first two roles. Greeks believed divine intervention on the part of the gods was responsible for flood waters were sent as punishment, or as a warning. Many cultures have stories of great floods sent by the gods. The Aztecs even sacrificed young children to be looked upon favorably by the rain god Tlaloc. In 1598 King James VI of Scotland is said to have believed that witches had summoned the gods of the wind to keep Ana of Denmark from reaching the shores of Scotland. Witchcraft executions related to severe weather peaked in Europe from 1560 to 1600 following decades of extreme weather, especially flooding. During the droughts of the 1890's the U.S. Congress began investing In rain-making experiments. In 1890, $9000 was spent to promote "concussion trials to literally "shake" rain from the sky. Many in the miliary already believed that the loud sounds from artillery were a positive force in creating rain. And, just this year, in 2023, Congress has agreed to spend $9.4M to advance the technology on cloud seeding.

Although we now accept that droughts occur periodically, and better understand the causes of such events, back in the late 1800s and early 1900s when drought was suffocating California, authorities were willing to ty almost anything, to spend more money than they would like, just to have some rain. Enter the Diviners. Broadly speaking diviners are individuals who foretell the future. When it comes to ending a drought, *diviners are individuals who claim to be able to make it rain.* In much of the world, indigenous people performed elaborate rituals to encourage rain. These rituals included offerings of sacrificed animals, sacrificial dance ceremonies and, in Thailand even a cat dance.

So, dances were passed down from generation to generation. Did they work? Did they make rain? Apparently enough time to sustain the rituals.

Diviners

Frank Melbourne was one of a small group of purported rainmakers active in the Great Plains of America during the 1890s. Melbourne practiced and popularized the idea of igniting gas on the ground as his primary means of creating rain. The concept was that the cloud created by the ignition ascended and interacted with air in the upper reaches of the atmosphere, causing rainfall. Actually Melbourne's theories were not radically different from the actual "evaporate/lift/condense/precipitate" model now understood to create precipitation.

Another rainmaker, Chares Hatfield, a self-professed "moisture accelerator" told San Diego's authorities in 1915 that he could fill Morena Reservoir, only one-third full, to overflowing within a year for a fee of $10,000, to be paid only if he succeeded. He proposed and constructed a contraption that consisted of a 20-foot high wooden tower built on stilts from which he released into the air a secret concoction of 23 chemicals. "I don't make rain," Hatfield said. "That would be an absurd claim! I simply attract clouds, and they do the rest." During Hatfield's work to wring water from the skies, San Diego experienced its wettest period in recorded history. Nearly 30" of rain fell in January and early February of 1916. So much rain fell that farms, homes and businesses were swept away. At least a dozen people drowned. Hatfield had to run for his life from angry farmers. Too much of a good thing is apparently worse than none at all. Critics claimed that he was a huckster who merely benefited from coincidence. However, in reality, by working only dry spells, the probability that rainfall would occur increased significantly. Hatfield also had an excellent understanding of meteorology, including weather fronts and low pressure systems that typically come ashore along the Pacific Coast during the winter months. He also had access to weather records. By knowing when storms were imminent, he could go where the rain was most likely to form. Although a rudimentary meteorological understanding of evaporation and condensation may have been part of their "bag of tricks". Most rain diviners were simply playing a hunch that atmospheric conditions were favorable for producing rain.

A Short History of the Roof

Since they first appeared on earth, humans have needed shelter from the rain; shelter to <u>keep themselves, their families</u>, their food, clothes and other belongings <u>dry</u>. Aspects of our lives we simply take for granted today. Early man started staying in a place that provided at least some protection –the caves. Cave fires provided heat, some limited protection from wild animals, and a place for cooking. Then sometime, several millions of years ago, during the Paleolithic Era, man moved out of the caves and into man-made shelters. These primitive structures were made of stones and tree branches. The stones were placed at the base of the structure to hold the branches in placed animal hides were stretched across at least a portion of "the roof" to afford some limited protection against the rain. Roof technology and application grew in step with building technology. Dirt was formed and dried into clay blocks, and before long bricks became the basic building block of shelters.

The Assyrians found that "firing" the bricks made them harder and more durable. Glazing those same bricks made them stronger and improved their imperviousness to water. With stronger foundations and structures "roof joists" could be laid across the top of the home and various materials could then be used to construct the actual roof itself. About 10,000 years ago in Mesopotamia, roofing materials were made from mud brick and mud plaster. The value of reeds and thatch in constructing roofing materials took hold. In the 7th century B.C., thatched reeds were used with mud, wood and stone for roofing.

Thatch became a common roofing material in ancient Greece. As skills improved in the manufacture of fired ceramic materials, roof tiles became increasingly prevalent. For this to work the walls had to be strong enough to bear the weight of the roof tiles. It was here, not necessarily in conjunction with roof tile weight, that the Romans significantly improved he arch with a combination of stronger cement, marble fired clay bricks and other enhancements. Meanwhile, in India, where home construction was undergoing surprising improvements, including stronger materials and instrumentation to ensure proper alignment of walls, home roofs for the first time were flat and made of wood. Although the Romans introduced tiled roofs to Great Britain around 100 BC, the British Isles were to become the home of the thatched roof. By 750 AD thatched roofs had become commonplace.

Thatch was a perfect solution for areas lacking in the typical inorganic materials used in other locations. Thatch was readily available, inexpensive and aesthetically pleasing. But thatched roofs also required significant maintenance, degraded with time, and were highly flammable. And many wild and domesticated animals became comfortable living in the thatched roofs. When it rained the animals often tried to wait it out, hoping the rain intensity would lessen or it would stop altogether. If in fact it started to rain heavily and the animals decided to look for alternative locations out of the rain, they began to jump from the roofs to the ground, leading to the famous saying,

"It's raining cats and dogs!"

Here is a summary of roofing materials currently available and in use around the world:

1. Asphalt Shingles – These are the default shingle for most American homes. They reliably provide a life of 15-20 years or more, and can be replaced easily though not necessarily inexpensively. They are available in a myriad of colors to match house and trim.

2. Wooden Shingles – Once the rage, the wooden shingle is increasingly losing favor with municipalities, builders and homeowners. Wooden shingles are expensive, difficult to repair and most importantly a fire hazard.

3. Tiles – Tiles are used predominantly in Europe. In Holland, for example, tile roofs are as prevalent as asphalt shingles are in the USA. The tiles are very heavy and home walls and foundations must be designed to carry such loads. Finding and repairing leaks in tile roofs is time-consuming and difficult.

4. Metal sheeting – Sheet metal roofs have a number of advantages. First of all, the use of large sheets of <u>impervious</u> metal minimizes the seams in the roof and therefore potential leak sites. Metal roofs may be made of aluminum or stainless steel and come in a number of colors. The main disadvantage of metal roofs in that <u>rain falling</u> onto them creates a loud and distracting sound, making sleep and even talk challenging. For this

reason, metal is used primarily as an accent, showing up more on for instance porch roofs than on roofs above bedrooms.

5. Flat roof composites and liquid coatings – Flat roofs have always presented challenges to builders and architects. Because water tends to linger much longer on flat roofs than on sloped surfaces, designers have had to resort to alternatives. This includes placing an imperious blanket of for instance ice shield under asphalt shingles, and coating especially tall city buildings, and single story buildings at a mall with a thick layer of hot liquid asphalt.

The Umbrella

There are few items used around the world for which it can be said such an item is ubiquitous. There may be others, but the one that comes to mind is the umbrella. For pure functionality without troublesome weight or size, nothing can out-do the umbrella as a first line of defense against the rain. Umbrellas have been around for as long as man has wanted to shield himself /herself from the rain. The umbrella was invented more than 4,000 years ago. It is also known as a parasol. The umbrella was first designed to provide shade from the sun. Improved water (rain) resistance was achieved by waxing and lacquering what were the first paper parasols.

Evidence of umbrella use has been confirmed through archaeological discoveries in Greece, Assyria, China and Egypt. "Umbrella" comes from the Latin word "umbra" meaning shade or shadow.

Umbrellas became popular first in northern Europe in the early 1800's, although in the beginning only for women. The first umbrella store opened in London in 1830. The first European umbrellas were made of wood or whalebone, and covered in mud. Steel-ribbed designs followed a few decades later. It took another decade for compact, collapsible, lightweight umbrellas to make their way into the marketplace. With the advent of the pocket umbrella in the 1920s, the umbrella industry continued to strive for the smallest, lightest, easiest to open umbrella. Over 3,000 active umbrella patents are registered with the U.S. Patent Office. Bucking the trend for small, and emphasizing converge instead, are the so-called "golf umbrellas" with diameters of up to five feet. About 33 million umbrellas are sold annually in the U.S. Plenty to keep us dry.

Are People Really Afraid of Rain?

As with many aspects of nature, humans have varied physical and emotional responses to rain. At the first drop of rain, many people will instinctively don jackets and hoods, and/or run for shelter. This is especially amusing when it happens at a warm weather resort and the individuals seem to be running for their lives. Of course, such behavior may occur to maintain a hairdo or make-up, or to avoid getting particular clothes wet. Such simple concerns are not generally classified as real fears.

But when just the thought of being rained upon is enough to keep a person homebound, a psychological disorder or phobia may be the cause. The technical name for the fear of being rained upon is pluviophobia, which could, for instance, have stemmed from a childhood trauma associated with flooding or landslides caused by heavy rains. An intense and irrational fear of rain itself, not just being rained upon is known as ombrophobia. Such a condition can include fear of major rain events, or something as minor as the fear of "catching a chill", after being exposed to damp weather. To those suffering from multiple days of cloudiness and precipitation, phobias of this type can lead to apathy and depression, portrayed well by The Carpenters' famous song "Rainy Days and Mondays Always Get Me Down".

Cloud Seeding

Cloud seeding has long held the promise of providing welcome rain in locations suffering protracted dry spells or drought. Although there are instances where rainfall increased after cloud seeding, many remain skeptical that there is a verifiable, repeatable relationship between seeding and rain. Cloud seeding essentially "encourages" the process of raindrop formation by providing additional nuclei around which water can condense. These nuclei can be salts, calcium chloride, dry ice, or silver iodide. Silver iodide is effective because its form is similar to ice crystals which are known to provide excellent sites for condensation and raindrop formation.

Many countries have for many years experimented with various chemicals using different forms of equipment to inject those chemicals into various levels of the atmosphere, gauging success on different criteria. The first known attempt for such a trial was attempted in Europe. And these and subsequent experiments, although looking primarily for rain creation, have also been used to reduce threats from hail in Canada, and to increase snowfalls at ski resorts. China has the largest standing research program in the world. Most are waiting to get past the creation of a few drops or flakes to a point where more continuous rainfall, not just wetting the road, but using the storm sewers can be expected with some regularity.

Acid Rain

Acid rain describes precipitation whose pH is well below neutral due to the presence of sulfuric and nitric acids formed when sulfur dioxide and nitrogen oxide gases interact with water in the atmosphere. Power plants release the majority of sulfur dioxide and much of the nitrogen oxides when they burn fossil fuels, especially coal, to produce electricity. The exhaust from cars, buses and trucks also releases significant quantities of sulfur dioxide and nitrogen oxides into the air. Other contributors are sulfuric dioxide from volcanoes, nitrogen oxides from lightning, soil bacteria and forest fires. Acid rain has been shown to have wide-ranging harmful effects on plants, aquatic animals and infrastructure. Included are adverse effects on forests, freshwater and soil, and as the cause of paint peeling off buildings, corrosion of steel support members on bridges and roads, and weathering of stone on buildings.

Rainbows

On the subject of rain, here is a brief story of the rainbows. Rainbows are truly one of nature's most beautiful spectacles. A rainbow is a luminous atmospheric arc featuring all the colors of the visible light spectrum. As its name would imply, rainbows can't exist without rain. If you see the sun come out during or just after a shower has passed you by, stand with your back to the sun and look for what will hopefully be an extraordinarily beautiful rainbow in the sky ahead of you. It won't always be there—conditions have to be just right for a rainbow to form; miss one piece of the puzzle and it's gone. Rainbows are formed by a complicated set of light phenomena as the sun shines through air containing water spray or raindrops. These light phenomena include refraction, refection and dispersion. Dispersion is the splitting of white light when it passes through a glass prism into its constituent spectrum of colors (violet, indigo, blue, green, yellow, orange and red). These are the colors we see in a rainbow.

Rain Headlines

- Rain is by far the most prevalent type of precipitation.
- Rain, even light rain, will send people scurrying, even at warm tropical resorts.
- Some games, like football, are played in the rain; others, like tennis, can hardly tolerate a drop.
- Water is the world's cleanser; when it evaporates, everything else is left behind.
- Some rain sensors in northern Chile have not registered a drop of rain in 100+ years.
- There is nothing like curling up with a good book on a rainy afternoon.
- Rainwater always flows downhill.
- Rain can transform a dried out lawn to lush green in just a day or two.
- Newer model year cars sense when it's raining and turn your windshield wipers on automatically.
- Rain is the bane of a convertible car owners.
- The word "rain" has found its way into America's many colloquialisms with "I think I will take a rain check for that", "Don't rain on my parade", and "Rain, rain go away, come again some other day!"
- Metal roofs are great until it rains.
- Rain can fall as quietly as drizzle and as noisy as wind-driven rain on your bedroom windows.

- Heavy rain can clean roadways of winter's leftover salt, sand and chemicals.
- Some people, who live in Earth's driest climates, may only experience rain once or twice in a lifetime.
- Others may have rain falling on them every day but a few, for the entire year.
- There are certainly hundreds, maybe thousands, of song lyrics that contain the word rain, the most identifiable of which is probably "Singin In The Rain".
- But also "Kentucky Rain" (Elvis); "Let It Rain" (Eric Clapton) and "I Wish it Would Rain Down" (Phil Collins).

If all the world's fresh water was collected and in turn spread evenly over every square of the Earth's surface, everyone would receive about 39" of rain water annually . Deserts are so dry because one critical component is missing from the "evaporate-lift-condense-precipitate" model for rainfall. By definition a desert receives no more than 10" of rain in a year

Rain's Contribution to Colloquialisms

- Our relationship with rain has led to many sayings.
    - o "April showers bring May flowers."
    - o "Come rain or come shine."
    - o It's raining, it's pouring, the Old Man is snoring."
- Some uses of the word rain become part of our colloquialisms, where we might not even realize that "rain" is in the saying
    - o "I'll take a rain check."
    - o "Don't rain on my parade."
    - o "You are invited to a baby shower."
    - o "Into every life a little rain must fall."
    - o "It never rains but it pours."
    - o "Rainy day fund."
    - o "Right as rain."

Final Thoughts

Rain is by far the most prevalent form of precipitation and an important component of the Water Cycle. A simple "Evaporate-Lift-Condense-Precipitate" model provides an easy to way to understand means for understanding how rainfall is produced in October. On the other side of this coin is the heavily weighted technical half. Some aspects of that technology include assessments of complex physical and meteorological concepts well beyond the scope of the average reader. These were the Ice Phase Process and Collision Coalescence Process, both of which have are based solely on theory. When too much water flows in too short a time, river flooding can occur, crippling surrounding towns. Flash flooding is actually the deadliest form of flooding. Too little rain can also be a problem, with drought developing slowly but surely often almost imperceptibly. Heat waves are similar to drought, just shorter; heat waves can hasten and/or aggravate existing drought. So, precipitation will only occur when water vapor is lifted significantly high in the atmosphere to allow condensation to occur. Although all four ingredients of the rainfall model (evaporation to produce the water vapor that rises into colder areas of the atmosphere where condensation occurs, and precipitation is produced) must be "working" for rain to occur. The mere presence of the ingredients does not mean it will rain.

CHAPTER 2 — SNOW

The processes that lead to individual snowflake formation are based on complicated physical laws governing crystallization that occur under conditions not seen to the same extent in other meteorological phenomena. Notwithstanding the complexity of the processes, the resulting product, the solitary snowflake, is a marvelous, beautiful, unique creation.

When many millions upon millions of these flakes band together and fall to the ground, a snowpack is formed. Not all snowflakes reach the surface. They may melt when they encounter an area of warmer air during their descent. Or a smaller percentage may undergo sublimation, which is the change in state of a material, in this case a snowflake, directly from solid to gas without first becoming a liquid. Be it a short-lived snow squall or a major snowstorm, the landscape will be covered by a white blanket, perhaps for only a very short time, or alternatively, depending on where the snowflakes fell, for the remainder of the winter season.

Ice crystals form in the atmosphere when extremely cold water comes in contact with particles of dust or dirt at a temperature below 32F, a state called supercooled. Nucleation of ice crystals is also enhanced when there is abundant water vapor available, a state called supersaturated. Ice crystal formation occurs most readily when the atmosphere is both supercooled and supersaturated. The tiny ice crystals that result are the seeds from which snowflakes will grow. Although snowflakes are most commonly found in nature as clusters of ice crystals, rare instances of "simple" snowflakes can occur. When the temperature is between 32F and minus 33F, the snowflakes are said to have resulted from a process called heterogeneous nucleation. Homogeneous nucleation takes place at still lower temperatures, below 33 degrees Fahrenheit.

Snowflakes grow as they fall through the atmosphere from 1) water vapor in the air freezing directly onto the ice crystals' surfaces and 2) ice crystals colliding with, and sticking to, other ice crystals. The growth of snowflakes, and the rate at which that growth occurs, is a delicate balance between water molecules moving slowly enough to stick, but fast enough to actually bump into individual crystals. Some snowflakes don't reach the ground, falling prey to melting and sublimation during their descent. Those snowflakes that do reach the ground may contribute to the formation of a snowpack. Although snowflakes are most commonly clusters of ice particles, there are rare instances of "simple" snowflakes. Probably most of the ice that comprises a snowflake is ether broken or a fragment of ice crystals.

Minerals are naturally occurring solids with a defined chemical composition and an ordered internal structure. Ice crystals are minerals. When super-cooled water vapor attaches itself to a particle, the ice crystal immediately begins the process of coalescing and arranging itself into an ordained crystalline pattern. In this case, the pattern is that of the distinctive hexagonal crystal of the snowflake. Nucleation encourages liquid $H2O$ molecules, upon freezing, to form crystal-like nuclei onto which other crystals can latch. Many water molecules find themselves sharing the same supercooled, supersaturated back-yard. Once close enough to be physically attracted to each other, the molecules move into a position where they can line up with their neighbors' charges.

Strong forces between the negatively charged and positively charged parts of adjacent particles leads them to join in a very specific three-dimensional pattern with six-sided symmetry. All snowflakes contain six sides or points. This arrangement allows water molecules to join and grow together in the most efficient way. As the snowflake travels through the water vapor-laden atmosphere, it picks up moisture which freezes and stays stuck. New crystals, each with the signature six arms of a snowflake, are formed. And so it goes, ad Infinium.

The levels of temperature, humidity, air currents, saturation, altitude, dust and dirt density and other high level and subtle variables <u>affect the size, shape and design of individual flakes</u>. For instance, colder temperatures usually lead to more intricate snowflakes; warmer temperatures produce smoother flakes with rounded edges.

So far no one has seen, let alone can show, that any two snowflakes are the same. Snowflakes come in millions of different shapes and sizes. The variables affecting snowflake formation further assure that diversity. High clouds produce six-sided hexagonal snowflakes. The average dendrite is comprised of about one quintillion (1 followed by 18 zeros!) individual water molecules. So statistically speaking, it really can never be proved using existing measurement methodologies that any two snowflakes are the same. Middle height snowflake production leads to flatter structures. Although individual flakes can be quite different from each other, the hexagonal manifestation is ubiquitous.

Wilson A. Bentley was the first person to take a picture of a snowflake. Bentley became an expert in the field of photomicrography, the photographing of very small objects. The addition of a quickly melting object to the process makes his achievement all the more remarkable, Bentley photographed more than 5000 snowflakes between 1885 and his death in 1931. He claimed that all of the snowflakes he observed or photographed were unique. A "Morphology Diagram" shows the shapes of snowflakes that form based on the temperature and degree of saturation of the atmosphere. Snowflakes are typically about one-half inch in diameter. Two-inch diameter flakes may occasionally be seen. Although perhaps folklore, witnesses purportedly corroborated an eyewitness report of a 15" diameter snowflake at Fort Kent, Montana on January 28, 1887. Other very large snowflakes were reported to have fallen in England that same month. Similar sightings were reported years later in Nashville, Tennessee and in England. Snowflakes can be classified into these six broad categories: dendrites, plates, columns, needles, irregular and rimed forms. Dendrites, from the Greek word "dendron" (tree), are snowflakes with symmetrical branches, six for each snowflake. with smaller branches coming off the main trunk. Other more specific types of snowflakes include sectored needles and rimmed crystals and spatial dendrites. Spatial dendrites prove snowflakes don't have to be flat to be snowflakes. Spatial dendrites are made from many individual ice crystals mixed together. The temperature needs to be below 32F for a snowflake to form, but not too much below 32F. If the air is too cold, there may not be sufficient moisture for condensation and precipitation to occur. While it can be too warm to snow, it can't be too cold, at least theoretically. Warmer air can hold more water vapor; the heaviest snowfalls typically occur when ambient air temperatures are at or above about 25F. This is why every winter we see (often large) snowflakes tumbling to earth when surface air temperatures are in the upper 30's. Snowflakes can survive at temperatures up to 41F, but no higher. Falling snow may eventually reach the ground in sufficient intensity for accumulation to occur. Snow varies remarkably depending on its water content. Wetter snows (with water contents off 10% or higher) are more likely to occur in the U.S. East; dryer snows (with perhaps 2% water content) occur more frequently in the U.S. West, where it is more likely to be called just "powder".

Snow is Quiet

There's nothing quite like the peacefulness one feels during a snowfall. Quiet, serene, tranquil. But it's not just "that feeling"; there are reasons for this quiet serenity. Snow has a pronounced effect on how, and how far, sound travels; the falling snowflakes themselves provide a barrier for noise transmission; the larger the flakes, and the heavier the fall, the more noise is muffled. The sounds you are used to hearing from your front porch or your bedroom, everything from airplane take-offs and landings to cars on the nearby, busy Interstate, to even the barking of a neighbor's dog, are suddenly neutralized. Second, snow, soft by nature, makes almost no noise when reaching the ground; no splatting of heavy rain on a metal roof or an asphalt parking lot. Third is the amazing noise absorptive qualities of freshly fallen snow. Noise absorption is measured on a scale of 0-1. Fallen snow has an absorptive factor of 0.6, equal more or less to industrial noise insulating materials like foams and fibers. As with those materials, freshly fallen snow has lots of air between the flakes. This empty space provides a home for noise—the snowpack takes in the sound waves but won't let them escape. With time, melting occurs, the snow refreezes at night, and the absorptive characteristics of freshly fallen snow are substantially reduced.

Snow is Not Slippery.

Snow is actually not slippery, which may be a surprise. In fact, at very low temperatures, snow is crunchy to walk on and a lot like sandpaper. Snow is much more slippery near the freezing point of water. When contacted by a warm tire or a warm shoe sole the immediate surface of the snow melts, creating a thin, molecular-like layer of water. Here the surface acts more like a liquid than a solid. It is this thin layer of water on the surface of snow that is in fact slippery, and potentially dangerous to pedestrians and drivers alike.

Snow is Reflective

Albedo is the term used to quantify the reflectance of the earth's surface. Snow has a high albedo, reflecting over 90% of the sunlight which it contacts. On the other hand, mountains devoid of snow as well as dark ground, plants and oceans have lower albedo. Albedo is a primary factor that determines the amount of sunlight absorbed into the climate system. The balance between the incoming sunlight and that directly reflected or absorbed helps determine the overall warming or cooling of the planet. Even small changes, such as lowered reflectivity caused by dirt, sand or pollution in ice, can have a net positive effect on albedo.

But major changes result when large areas of land transition from snow-covered to bare dirt and rock. This process is referred to as "feedback". One can create this phenomenon on a snow-covered asphalt driveway. Dig a hole in the snow about one foot square so the sun can shine directly on the asphalt. Notice how quickly the hole grows in size as the asphalt absorbs more heat than the white snow.

Snow Melts

Well. It really can't last forever. Except in the coldest reaches of the Earth where temperatures rarely go far above freezing, all those square miles on which snow rested in mid-winter, are barren by June or July.

A very small percentage of snow melts occur from the bottom up. Soil temperatures, shielded from often very cold air temperatures above the snowpack, may be just warm enough to initiate a little melting. Another very small contributor is sublimation, the process by which solid snow undergoes a phase change directly to water vapor. A snowpack melts predominantly from the top down. Perhaps surprisingly, neither sunshine alone nor rain alone are predominant factors affecting the rate of snowmelt. But wind and humid air are. <u>Sunshine working together with warm, humid winds create the maximum melting conditions</u>. Similarly <u>rain and warm wind</u> cause rapid melting. Some lead time is required before melted snow begins to flow. In the early phases of melting, much of the water remains in the snowpack. It's not until the snowpack contains about 10% liquid water by weight that it reaches its holding capacity and water begins to flow.

A July 1998 technical paper by Linacre and Geerts describes processes that contribute to snow melt. They ranked their approximate order of importance, depending on local conditions. The first four processes are:

1. Net incoming solar radiation: the older the snow cover, the lower its albedo, and the more solar radiation it absorbs;

2. Air: at the same air temperature (above freezing), snow melts faster in windy conditions than under calm conditions;

3. Terrestrial radiation emitted by the clouds and sky (sky radiation): the emissivity of snow is very close to one, therefore it absorbs almost all incoming infrared radiation (in fact it is the infrared component in the solar radiation that explains why snow melts under bright sunlight when the air temperature is just below freezing);

4. Heat transfer from the ground below: the snowfall associated with a cold snap following a warm spell is less likely to last long.

In addition, we can add these considerations

- Snow's Cleanliness – A large expanse of unbroken white snow effectively reflects much of the Sun's energy. Any dark-colored contaminant such as even a tiny bit of dirt, deposited on the snow's surface, will absorb heat from the Sun and accelerate melting.

- Wind - Much like blowing your breath on hot soup cools it down, a warm breeze accelerates the rate of snow melt by continuously sweeping away the boundary layer at the snow/air surface and resupplying that interface with warm air.

- Rain – Heavy rain, and especially heavy warm rain, melts snow by the especially effective direct heat exchange characteristics of liquids. Put an ice cube in the sink. Even if the room temperature is 68 degrees Fahrenheit, it takes quite a while to melt. Now, drip cold water (say 40F) from a spigot, and see how quickly it disappears.
- Humid Air – When the dew point is high, vapor from the milder air is deposited onto the colder snow surface, hastening melting.

With so many factors at work melting snow, and with so much complexity regarding the characteristics of snow awaiting melting, it has (not surprisingly) proven very difficult to develop an algorithm that can predict snowmelt rates. Practically speaking, snow depths of up to six inches can melt in one day. Models suggest that up to 20" a day are possible under perfect melting conditions. Rates that high were reported in the mid-Atlantic states following snowstorm of 1996. And there is a documented case in Alaska in which nearly six feet of snow disappeared within two days when it was literally eaten by 50F Chinook winds.

Measuring the Depth of Fallen Snow

Snow depth is the single most difficult weather parameter measurement, easily surpassing measurements of temperature, humidity, and wind speed and directions.

How does the National Weather Service measure snowfall? Pick a location shielded from the wind but not so shielded that some snow will not be collected. "Catch" the snow on a level whiteboard, which is nothing more than a four-foot square by two-inch thick piece of plywood painted white. Measure the depth of the fallen snow with an ordinary ruler. Clean any snow off the whiteboard every 6 hours to minimize fallen snow compression, melting and wind effects. This all sounds quite easy, but the reality is that a number of features and factors, some human induced, conspire to create obstacles to creating a consistent, repeatable snowfall measurement. Since snow is carried by the wind, a measurement taken on your deck might be greatly enhanced by snow that has blown off your roof and onto the deck. High winds might remove snow cover from a nice spot in your backyard you thought was perfect for measuring fallen snow. Snow compresses continually as soon as the first flake covers the ground. After 10" of snow has fallen and sat for a while, the depth will be more like nine inches.

Rain and/or warmer weather conditions may occur during the National Weather Service's recommended six-hour measurement window. Such a change in ambient conditions will easily eat away heavy snowfalls, and can wash/wipe away smaller accumulations completely, as if the initial snow event never happened. If the observer is certain that the snowpack on a snowboard is likely to diminish or disappear due to rain or warmer temperatures, he/she can then take a special observation before that melting occurs and report that total.

High tech acoustic sensors, radar, microwave and satellite imagery are all available for measuring snow depth, but are costly to buy and operate. These sophisticated tools are more typically used to define macro-scale issues like flood potential and water resource management.

The National Weather Service's advice to the general public for this difficult task is to measure fallen snow is to, what else, do it the way they do it. Mostly just try to emulate the wind protection/cold surface/periodic wipe/trusty ruler/honest reporting approach they use themselves.

To provide the most complete weather report, a measurement of the so-called "liquid equivalent" must be made. Liquid equivalent is the amount of liquid produced when snow or sleet from a frozen precipitation event is melted. This is important because the water collected directly from a warm summer rain event and the liquid equivalent from a winter snowstorm are of equal importance when calculating official monthly and annual precipitation reports. To determine the liquid equivalent, catch the frozen precipitation in the standard rain collector with its innards removed. When the frozen precipitation has ended melt the snow/sleet. Pour the water into a graduated cylinder and note the amount. This is the Liquid Equivalent.

Man and Snow - Recreation

Snow provides many enjoyable recreational activities—downhill skiing, sledding, snow-shoeing and cross-country skiing. But one need not move fast to have fun with snow. There are also recreational activities in the snow that don't require skis like building a snowman or a snow fort, or following animal tracks in the snow. Just make sure the tracks you are following don't belong to an animal with whom you would prefer not to "interact".

There are over 500 ski resorts in the United States, ranging from states like New York, which has around 50, to the states that have only one, like Maryland. There are also global destinations like Vail and Aspen. Park City Utah is the biggest U.S. ski resort. If snow has been plentiful at operations are at full capacity, Park City can provide 25 miles of trails. Artificial snow making has become more and more prevalent and more and more effective, allowing resorts to make up the difference in snow lost through warmer winters, undoubtedly saving more than a few from bankruptcy. Nonetheless there is an increasing risk that the least well situated and profitable resorts will be forced to go bankrupt. Less snow and higher temperatures have slowly eroded the ability of ski operators in the U.S. Mid-Atlantic from Pennsylvania to West Virginia to stay open.

Skiing is an important industry in the mountainous communities of the western United States. Over time, the sport of skiing has become more and more expensive as the costs of liability insurance and sophisticated snow-making equipment have increased dramatically. For the privilege of taking on the slopes, skiers get to spend for lift tickets and equipment rentals (prices may vary, and discounted lift tickets are often available). Even so, you can expect to pay upwards of, or maybe more than, $100 per day or more for lift tickets and equipment rentals.

Snow Around the World

Snow is found predominantly in the mountainous regions of the Northern and Southern Hemispheres. Annually snow covers about 18,000,000 square miles of North America, Greenland, Europe and RUSSIA combine.

Snow can also occur in locations close to the equator which are generally thought to be too warm for snow. Like Hawaii. The two primary factors that determine whether snow will fall, and stick, in the northern hemisphere are latitude and altitude.

- The greater the latitude, the more likely it will snow, all other factors remaining equal.
- The higher the altitude, the more likely it will snow, all other factors remaining equal.

Despite being in a latitude-favorable position, Reykjavik, Iceland receives very little snow because of the Gulf Stream. Other factors that contribute to snow formation and retention include

- The amount of snow that falls during the winter
- The distance to the nearest lake or ocean. The water remain cooler than locations surrounded by land, which heats up more quickly and retains that heat longer
- The relative location of the sun
- The dryness of the atmosphere
    - o In locations like Antarctica, the humidity is so low that minimal snow falls even when latitude and altitude favor its formation.

Few places on earth consistently receive more snow than the western coastal mountain ranges of North America, including the Cascades of Washington and Oregon and the Sierra Nevada of California, as numerous sites with very high annual snowfalls, including Paradise Lodge in Mount Rainier National Park in Washington State, and Crater Lake National Park in Oregon. Other locations with prodigious annual snow totals include the so-called Japanese Alps that cross the island of Honshu, and the coastal mountain ranges of Alaska. It's the mountainous terrain and the effects of orographic precipitation that are responsible for these heavy snow totals.

Many of these mountains are in <u>remote locations</u> and snow depths are typically estimated using visual observations and images from satellites. The process of validating annual snow depths at locations around the world invariably leads to confusion, contradictions and uncertainty.

The United States is home to three of the five snowiest places on earth, starting with Thompson Pass, Alaska, which regularly receives more than

550" of snow each year; some years it is estimated that as much as 900" has fallen, The U.S. Inter-Mountain West regularly has numerous sites with very high annual snowfalls. Aside from the Pacific states, the snowiest places in the Wasatch and San Juan Mountains of Colorado, which receive over 400" of snow per year. Japan presents an especially difficult challenge for those who measure snow depth. Honshu Island, Japan is the home of Japan's snowiest mountains. But finding consistency in the snowiest locations in Japan, let alone the actual depths of snow measured at each location, has proven difficult. Recently reported annual snowfalls from Japan include Aomori (310") and Takada (262"). The greatest snow depth ever measured on Earth, 465", was recorded on Mt. Ibuki, Honshu, Japan on February 14, 1927. Not to be slighted, Europe is represented in fifth place by the Alps of France and Switzerland. The Swiss Alps average a respectable 440", and of course there are a lot of additional non-USA locations like Norway, Quebec and Newfoundland that receive significant annual snowfalls.

Snow in America

Here is an interesting tabular summary of the annual snowfall in America's cities with populations greater than 50,000. These totals represent their 30-yeasr averages for the timeframe 1991-2020. This important metric provides the best representation of snow depth (or for that matter any meteorological parameter – temperature, rain). Thirty years is enough to smooth out or fill in the exceptional record highs or lows. Seems like it wouldn't, but it does. It was conceived in the 1930's and continually endorsed by the World Meteorological Organocation (WMO).

Location	Average Annual Snowfall
	Inches
Syracuse, NY	123.8
Flagstaff, AZ	101.7
Erie, PA	100.9
Rochester, NY	99.5
Buffalo, NY	94.7
Boulder, CO	88.3
Duluth, MN	86.1
Grand Rapids, MI	74.9
Cleveland, OH	68.1
+South Bend, IN	66.6
Pocatello, ID	66.4
Worcester, MA	64.1
Great Falls, MO	63.5
Portland, ME	61.9
Chyanne, WY	60.3
Albany, NY	59.1
Battle Creek, MI	58.1
Fargo, ND	57.9
Ann Arbor, MI	57.3
Provo, UT	57.1
Billings, MT	55.0
Denver, CO	53.8

Of course, many smaller cities have impressive annual snow totals, like Valdez, AZ coming in with 300+ inches of snow and a population a little more than less 5,000 and Truckee, CA, population less than 18,000, sporting an average annual snowfall just a smidgen more than 200" And if look simply at U.S. "locations", rather than cities, we get a markedly different list (this one for the winter of 2021-2022).

Location	Average Annual Snowfall (")
Paradise Rainier Ranger Station, WA	680
Thompson Pass, AK	550+
Mount Baker Lodge, WA	530
Crater Lake, OR	530
Alta, UT	516
Soda Springs, CA	471
Tamarack, CA	445
Stampede Pass, WA	442
Wolf Creek Pass, CO	436
Silver Lake, UT	429

Over the past hundred years or so, there have been instances where snow fell very rapidly, where it accumulated very fast and where the resulting total snow depth was just short of unbelievable. These events occurred in timeframes ranging from as long as an entire winter season to as short as one hour.

Where	Inches	Timeframe	When
Mt Rainier, WA	1,224.5	365 Days	2/19/71 to 2/18/72
Mt Baker, WA	1140	Winter Season	1998 – 1999
Valdez, AK	975	Winter Season	1952 – 1953
Thompson Pass, AK	974.1	Winter Season	1952 – 1953
Squaw Valley, CA	728	Winter Season	2016 – 2017
Valdez, AK	346	One Month	December 1955
Mt Shasta, CA	189	One Week	February 13-19, 1959
Valdez, AK	175	Five Days	December, 1955
Silver Lake, CO	95	32 Hours	April 14-15, 1921
Valdez, AK	62	One Day	December 29, 1955
Thompson Pass, AK	78	24 Hours	December 8, 2017
Donner Summit, CA	67	One Day	January 2, 1982
Oswego, NY	17.5	Two Hours	January 26, 1972
Copenhagen, NY	12.0	One Hour	December 2, 1966

We are generally most interested in who receives the most snow in a given time period. Here by comparison are the major cities in the U.S. with the lowest average for the five-year period 2016-2021.

City	Five-Year Average Inches
Albuquerque, NM	4.90
Medford, OR	4.68
Little Rock, AR	4.44
Tulsa, OK	4.36
Raleigh, NC	4.16
Memphis, TN	3.12
Knoxville, TN	2.72
Fort Smith, AR	2.00
Charlotte, NC	1.72
Nashville, TN	0.98

Building a Snowman – From an Engineer's Perspective

People have been building snowmen for centuries, certainly well before 1380 A.D. where history documents that the Belgians built over a hundred snowmen every year as part of a religious celebration. Building a snowman somewhere in your front yard or in a nearby park is a fun activity. But once you have decided to build a snowman three or more feet tall, you need to consider the characteristics of snow, and to understand some subtle but nonetheless important engineering principles. Otherwise, your snowman is doomed. So here is probably the most technical description you have ever seen to describe the process of building a snowman. The water content of snow is the single most important variable in the entire snowman-making process. Wet snow, common in the East and at lower elevations in the West, is by far the best for making snowmen—it sticks together and it's easy to roll into a ball. But wet snow is heavy, both to roll and to lift, two necessary steps in the building process. To overcome the difficulty in rolling snowballs made of heavy snow, let gravity help you by rolling the snowballs down a hill. Hopefully, it ends up in a location you like because chances are you won't be able to get the ball back up the hill. Fortunately, the opposite is also true; making a snowman from dry fluffy snow is next to impossible. Light snow doesn't stick and any snowballs that do get made from such snow are prone to cracking and general lack of integrity.

A snowman typically consists of three snowballs, one on top of the other. Once you decide where you want the snowman to be, you will have to create the next two snowballs. Enough about snow's characteristics. Now the assembly. You will have to lift the second snowball up and onto the first snowball and then of course the third up and onto the second. As alluded to earlier, these steps may prove exceedingly difficult for several reasons: 1) you can't get your arms around the second and third snowballs to lift them, and there's no contraptions to make it any easier and 2) you can't keep the growing snowman stable as the second and third snowballs are lifted into place, potentially causing snowballs to crack, imperiling the entire project. The second (and third) snowballs are so heavy you can't lift them. If any of the snowballs are made out of dry fluffy snow it may well break into pieces during the lifting. Or that same snowball made of dry snow may not be sufficiently strong to hold the weight of the snowball placed on top of it all three snowballs have been positioned it's time to decorate the snowman—nose, ears, eyes, mouth, hat, belt. From the classic "Frosty the Snowman" we know that Frosty had "a button nose, a corn cob pipe, and two eyes made out of coal". Today all kinds of paraphernalia are available to create a head that meets the needs of the builders and the snowman itself. Certain religious sects have even attached symbolism to specific aspects of the otherwise simple snowmen, including its white color, circular shape and carrot nose to signify such tenets as everlasting life and forgiveness of sin. But, for most of us, the snowman is just a likeness of ourselves that we can have as friends for a few days. Perhaps a happy smile on its face. It will stand as the snow around it melts, exposing the ground on which it rests. And eventually it too must melt, becoming unrecognizable as it slowly fades away.

Snow Removal

Early settlers, with little need to travel very far from their homesteads during the winter, generally viewed snow as a source of water for drinking and cooking, as well as a means to improve soil moisture for the coming spring's crops and livestock. As towns and cities grew, it was no longer sufficient to simply enjoy the beauty and tranquility of freshly fallen snow. Snow was becoming a problem.

Snow shovels were man's first weapon against fallen snow. Remains of what may have been some of the first snow shovels were found in Russia and estimated to be about 6,000 years old. It appears that the shovel's blade was made of an elk antler, and its handle of wood or bone. Today's snow shovels are not remarkably different from ones used thousands of years ago. After all every shovel needs to have a handle and a scoop. Over centuries, technology has been utilized to make it easier to push, lift and throw snow with a hand-held snow shovel. Fiberglass and certain plastics have slippery surfaces to which snow has trouble sticking, making snow tossing more effortless. Long handles are better for pushing snow, shorter shovels (especially with bent handles) for tossing. In the end, it seems there is no perfect design. More than a hundred patents have been granted for snow shovel designs since the 1870s. Shoveling snow is hard work. It's easy to strain muscles in your back, shoulders and wrist when lifting and tossing snow. Any given snow event may require a person to lift tons of snow at literally a moment's notice. So it is not unusual for snow shoveling to cause heart attacks. Slips, trips and falls are also common.

Until the onset of the automobile, horse-drawn wagons were the primary form of transportation. Ski-like runners provided smooth, consistent transport in the winter, provided of course there was enough snow to keep the roads snow-covered. In a strange twist from current priorities the goal was actually to keep roads snow-covered. Snow as purposefully added to roads and bridges. The snow was then compacted and flattened by a vehicle called a snow roller, essentially a giant wheel weighed down with rocks and pulled by oxen or horses.

For such snowfalls, or even ones with lighter accumulations, horses had to be replaced with machinery that could provide sufficient power to move significant amounts of snow.

Things were happening quickly in the world of "snow management". Although a large segment of northern America was still very rural and farmers had little to be concerned about from a snow event, in fact generally welcoming the moisture. But America was quickly becoming more urban, with cities like New York and Philadelphia growing fast. It just took one defining 3-day period for "those in charge" to realize that the current means of propulsion for removing snow, namely the horse, was just plain not going to cut it when it came to snow removal from New York City's streets. That event was the Blizzard of 1888.

In the early 1920s, two Norwegian brothers produced designs for vehicle-mounted snowplows. These were the forerunners of the plows of today that proliferate each fall and especially when snow is in the forecast, and someone might need a driveway or parking lot plowed. Companies and individuals continued to look for better ways to move snow more effectively, faster and cheaper. Although clearing snow from roads to allow cars easier passage was important, it was really for the railroads of the American and Canadian West that "snow blowers" were used to clear snow from tracks at high elevations. For it was here, in these mountainous regions, that snow fell often and periodically with prodigious amounts, measured as easily in yards as in feet. Machines were needed that could lift and throw snow far enough away from the immediate site to be cleared to allow trains to continue on their journeys. Enter the rotary snowplow, or as we know it today, the snow blower.

J.W. Elliott, a Toronto dentist, had designed a rotary engine that drove a wheel rimmed with flat blades.

As the plow went down the track, snow collected in a housing on the plow and then got funneled up to the blades, which tossed the snow out through an opening at the top of the contraption.

The plow cleared the track easily, tossing snow as far as 200 feet. Railroad managers were impressed; the plows were purchased and put to work. With time, snowblowers got cheaper, smaller, and easier to use. Truck-mounted models were designed for commercial enterprises. Eventually, snow blowers became small enough to be fit for home use. Shoveling snow became less and less common, reserved mostly for small driveways or light snow, where the shovel still worked well.

Today's Snow Clearing Machinery

Many state departments of transportation are increasingly utilizing, or strongly considering, the monster contraptions being used in snow-prone locations. Some of today's snowplows are huge, with multiple blades and the capacity to simultaneously plow and/or blow snow and treat roads with salt and sand. Departments of transportation and airport authorities in traditionally snowy parts of the country have found that these gargantuan pieces of machinery, particularly when deployed in a convoy, can clear even significant snow totals from roadways and runways quickly and efficiently.

Snow and Christmas have always been irrevocably tethered (at least in the Northern Hemisphere). This association has grown from the beauty and tranquility that are the essence of both the holiday and the precipitation. These thoughts were turned to song and memorialized in the now universally famous "White Christmas", sung best by Bing Crosby.

Location-Specific Probabilities of a White Christmas for Select U.S. Cities.

City	Probability		City	Probability
Fairbanks	100		Chicago	40
Duluth	97		Pittsburgh	33
Marquette MI	90		Boston	23
Portland, ME	83		New York	22
Spokane, WA	70		Philadelphia	15
Cleveland	60		Washington, D.C.	10
Hartford	57		Los Angeles	1

A White Christmas is declared if one or more inches of snow is on the ground at 7 a.m. Christmas morning. So it need not actually snow for a White Christmas to be declared. Snow from a storm days prior may provide sufficient accumulation to make the grade.

Believe it or not, some of the world's warmest regions have experienced winter blizzards, spring flurries, and even summer snowfall. Here are a few places where, quite surprisingly, snow has fallen.

Snow in Unusual U.S. Locations

Where/When	*Snow Event*	*Prospective*
Los Angeles, CA January, 1949	Half an inch of snow downtown Nearly a foot in the San Fernando Valley	Greatest recorded winter storm in city's history Has had measurable snow six times since 1949
Mauna Key, Hawaii June 14, 2016	Light dusting of snow	At 14,000 feet above sea level Mauna Key regularly experiences snow, ice and very cold temperatures . But in summertime snow is very unusual
South Florida Early morning of January 9, 1977	A trace of snow fell on Miami's beaches and snow flurries were reported at Homestead	First time snow had ever been reported this far south on the Florida Peninsula
Phoenix, Arizona January 20, 1933 January 21-22, 1937	An inch of snow each event	Snow is very unusual in Phoenix Ninety years or so later, these two snow events in the same decade remain as records. Since that time only a trace of snow has been recorded, that in 1998

Drifting is especially prevalent in the Plains states of the U.S. where snow is usually light in weight, temperatures are bitterly cold, the terrain flat and winds blow unimpeded for miles. Snow can drift considerably, completely altering the landscape. A snowdrift is, stated simply, a deposit of snow sculpted by wind into unusual shapes. Sometimes even a relatively meager snowfall can be transformed into the most unique curves and geometries—the ground itself is laid bare while huge collections of snow carried from those locations bury nearby homes, cars, and outbuildings. Snow will continue moving along a flat stretch of land or paved parking lots until it encounters an obstacle. The obstacle deflects air flow. The shape, size and position of the obstacle will determine the shape and size of the resulting snowdrift. The type of snow crystal and the wind speed determine when snow will begin to move. to move. The speed necessary to carry the lightest snow is about 11-12 mph. Sand dunes resemble snow drifts and are formed in a similar manner. Despite their beauty, snowdrifts present many risks to drivers, skiers, pedestrians, and homeowners. These risks include roofs collapsing, and avalanches caused by the additional weight of the snow collected in drifts. Snow drifts present a particularly high risk to drivers when they form on highways. The driver, caught in a sea of white, cannot distinguish the drifts from the road surface. The drifted snow quickly brings vehicles to a stop, creating the opportunity for multi-car pileups.

<u>Snow Fences</u>

Snow fences help keep snow from affecting specific features of the landscape, such as homes and roads. Snow fences were first built in Norway in the middle of the nineteenth century. Their purpose was to collect snow as a source of water for cooking and cleaning. Americans began experimenting with, and eventually using, snow fences in the 1870s to reduce drifting of snow in locations like the front door of a house or a barn. Snow fence design has become a science of its own. Engineers have debated, designed and tested snow fences of varying heights, materials and open areas virtually since the first fence was deployed. Regardless of the exact design features, all snow fences work by reducing the wind speed, causing the blowing snow to fall behind the barrier. A major use of snow fencing is on rural highways. Snow fencing allows State Departments of Transportation, not Mother Nature, to dictate where snow will collect.

That is, before the snow reaches the road, not on it. Since snow fences are used only in the winter and may be deployed on farmland, the preferred snow fencing is often the relatively light weight type that can be rolled up and stored as spring arrives. Trees, shrubs, and bushes can also act as "snow fences" by slowing the speed of blowing snow. Nature's snow fencing has the added benefit of no cost and no seasonal storage.

Snow avalanches are the rapid downslope movement of snow, ice, and associated debris such as rocks and vegetation. Avalanches can easily bury and potentially kill anyone in their paths. There are also avalanches that contain no snow. From a technical perspective snow avalanches can only occur when a cohesive slab of snow, lying on a weaker layer of snow, fractures and slides down a steep slope. The snow avalanche can be triggered by several micro- and macro processes from lab intra-slab snow mechanics to literally "the snowflake that broke the camel's back". Once they get rolling snow avalanches accelerate and grow as they entrain additional snow. The <u>forces</u> generated by moderate to large avalanches can damage and destroy most manmade structures. Trains crossing the Sierra Nevada mountains were often subjected to snowstorms that buried the train for days. Donner Summit averages about 35-40 feet of snow a year. That total has reached as much as 60-70 feet on occasion. Increasingly powerful and effective snow clearing equipment such as plows and blowers provided an improved defense against such periodic snowfalls.

The Snow Shed

But it was still difficult to use this equipment to handle the snowiest weather or an avalanche in the snow shed. The primary function of a snow shed was to keep snow off the train tracks in the first place. Snow sheds were generally enclosures made of pine. In essence they were tunnels. Twenty-three miles of snow sheds were built from 1867-1868 in the California mountains. Ultimately forty miles were constructed. Snow sheds were initially deemed a great success. However, despite the greater assurance of on-time, or close to on-time, arrival in Los Angeles, problems soon developed with the snow sheds and those problems multiplied. Passengers spent the time required to pass through the sheds in total darkness, sometimes accompanied by smoke from the locomotive. Passengers could not see some of the most beautiful panoramas in the world while inside the shed, winter or summer. The wooden sheds baked in the summer sun, attracting sparks from the trains' coal-fired engines or nearby forest fires. Eventually, in the 1960s, most of the sheds were either re-moved or replaced with fire-resistant concrete. Today's use of huge, powerful train plows and blowers keeps the train tracks generally free of snow and open for travel except under the most extraordinary snow conditions.

In some parts of America, winter can hang on for an excruciating long time, well into spring in many parts of the country. Place like Salt Lake City and Denver make this milestone. The last measurable snow in Salt Lake City falls ridiculously late (almost May 1). But many other cities are not far behind with even Boston and Pittsburgh waiting until sometime in April. Manmade snow piles can last even longer. There were piles of dirty snow at the Minneapolis - St. Paul Airport (MSP) in early July 2023. Even in the East, not known for the late season snows expended and extended winter seasons, Chicago porches an average last meas-urable snowfall of April 2, Detroit April 4 and Minneapolis April 8.

When Does Winter End?

Day of last measurable snow, July 1964 - June 2014

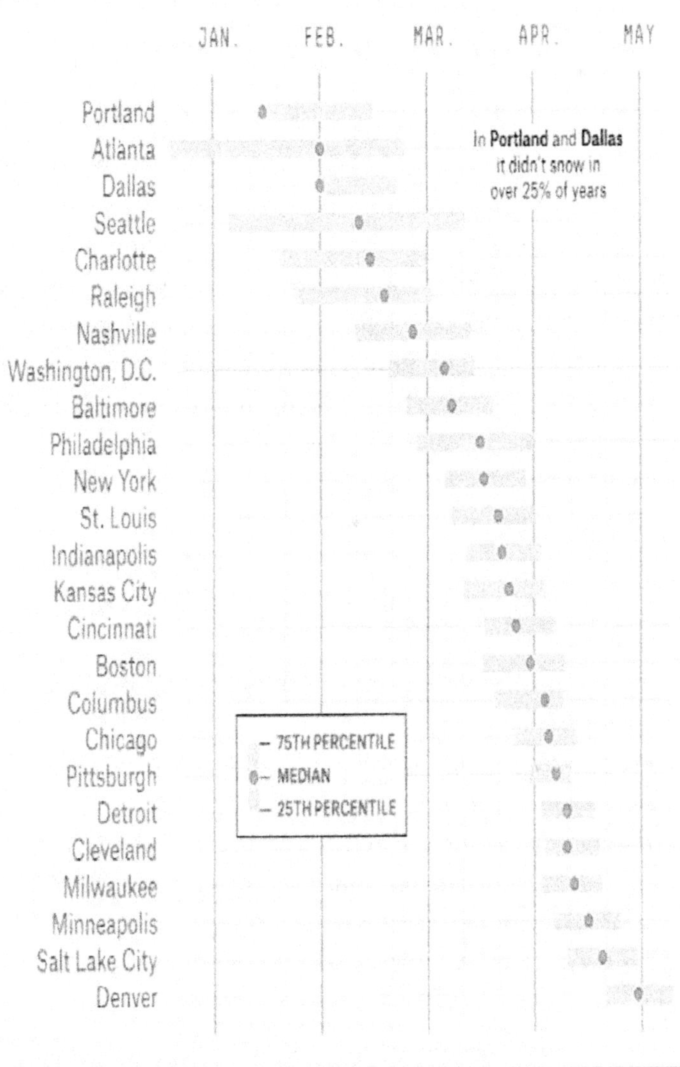

CHAPTER 3 – WIND

Can an article about wind really be that interesting or informative? How complicated could it be? After all, wind is just moving air! As turns out, there really is a lot to say about it. The wind is omnipresent, multi-purpose, essential, diverse, mysterious, helpful, destructive and intriguing. For instance:

- Wind is, by itself, invisible—you can't see it, you can't hold it.
- Wind is moving air that creates forces that becomes apparent when you see leaves moving on a tree, feel a chill on your face, or witness destruction from a tornado.
- Wind exists in the troposphere in the form of jet streams that play a major role in creating weather on the Earth's surface.
- Wind originates from uneven heating of the of the Earth's surfaces, and subsequent thermal and pressure differences between high and low pressure systems.
- Wind at the Earth's surface ranges from gentle zephyrs to startingly destructive storms.
- Wind is the Earth's great transporter, distributor and mixer of all things it can lift and carry, solids, liquids and vapors alike.
- Wind inspired the creation of gods and other mythological figures representing the directions of the wind, and which led to stories of their great powers, true or otherwise.
- Wind has been put to use by man over the centuries to propel boats and ships in times of peace and war, to create forces sufficient to do work like lifting water, milling grains and cutting wood.
- Wind technology eventually became capable of generating sufficient energy to light homes and power machinery.

- Wind is currently an expanding alternative energy source that can displace fossil fuels for some or many of man's energy needs.
- Wind is essential for the development and intensification of hurricanes and tornadoes, but can also lead to their demise.
- Wind of sufficient force can cause catastrophic damage to buildings and vegetation.
- Wind can knock down trees and tree limbs, taking overhead lines with them, leading to power outages that can last for days. Winds play a simple a major role in determining weather and climate by carrying heat, cold, water vapor, dust, volcanic ash and other transportable items to locations around the globe creates waves on bodies of water that intensify storm surges make people sick, and cause seashore beach erosion. Winds help drive global ocean currents.
- Wind carrying erosive sand particles can, over millions of years, create natural structures.

Wind Generation

Wind originates from uneven heating of the Earth's surfaces. Uneven heating is as important to the understanding of onshore and offshore breezes as it is to is to the development of ocean currents going thousands of miles. Uneven heating results from a myriad of geological, atmospheric and oceanographic features—sun intensity, cloud cover, percentage of earth covered by water, latitude and longitude, nature of land use—grasslands farmlands, forests, deserts, etc. On hot summer days the scenario goes something like this. As the sun rises, the Earth absorbs more of the sun's energy than does the water. The differential in heating leads to lower pressure over land which in turn encourages the resulting warmer, less humid air over land to rise. The pressure gradient then forces the movement of cooler, more dense air over water surfaces to move onshore to replace the air rising from the land. Onshore breezes provide cooling relief to beachgoers. It is often referred to as a sea breeze, As the sun loses its power in the late afternoon, the land cools more rapidly than the water, and the air adjacent to the shore reverses direction and flows outward toward the sea. Onshore breezes also create a zone of convergence further inland that can result in development of showers and thunderstorms along the sea breeze front! Those new storms can then move back toward the beaches and create dangerous conditions for beachgoers as a result crops of cloud to ground lightning, heavy rain, hail and high winds.

Jet Streams and Weather

Jet streams are fast-flowing ribbons of air and water vapor that are located high above the Earth's surface. There are two distinct jet streams, the Polar Jet Stream and the Sub-Tropical Jet Stream. It is the Polar Jet Stream that is of most interest to us. As it meanders around the Earth, the polar jet often intrudes on the meteorologically active 30-60 degree mid-latitudes, where it can have a major impact on weather and aviation. The polar jet stream carries weather systems with it as it moves around the globe. Without this pushing, pulling, and nudging weather would change little day-to-day at a given location. The polar jet is three miles deep (thick) and more or less 100 miles wide. The polar jet stream also undergoes occasional northward bulges and southward plunges. Meteorologically these bulges are called ridges (or high pressure systems) and troughs (or low pressure systems). So, the path of the polar jet stream directly affects high and low pressure system development, intensification and movement at lower levels in the atmosphere and on the Earth's surface. Understanding this relationship is an essential factor in forecasting the weather. During the warmer summer months, the polar jet, free from sharp temperature contrasts and resulting pressure gradient forces, retreats north into Canada. As the weather begins to get colder at the poles and temperature and pressure gradients become more pronounced, the polar jet begins a southward migration. It then, once more, becomes a major meteorological force. The second major jet stream is the Sub-Tropical Jet, which is also circling the globe, but at an altitude above 30, 000 to 39,000 feet, where weather occurs.

Many winter seasons, low pressure systems make their way up the Atlantic Coastline spreading rain and snow across wide areas of the Mid-Atlantic. First "discovered" by Benjamin Franklin in the 1700s, it is the so-called Nor'easters that are commonly associated with these storms. As they form and begin to move, these low pressure systems are responding to a number of critical atmospheric processes occurring on various zonal and vertical scales, including not only the pattern of upper level winds, but a whole host of other factors. These include diabatic heating and latent heat release from clouds/precipitation associated with the low pressure system. It's all highly non-linear; the low is responding to the sum of all of these processes (too intricate to be discussed here) and the track and intensity/deepening of the associated low is not merely following the jet stream. National Weather Service models are still not getting all of those processes modeled correctly; so this is (and will be) a work in progress (even as strides are made by the NWS).

Wind transports heat from the tropics and vice versa. Contributing to this process are, among others, the Hadley, Ferrell and Polar wind cells, and Rossby waves. These are complex concepts. The cells result from heat transport poleward from the tropical latitudes, but are strongly influenced by the earth's rotation, embodied in Coriolis forces. Other effects, like heat and momentum fluxes, actually control these cells. Overall, the scope of this segment is beyond the technical competency, or interest of most readers.

Wind Zones

The Earth has five major wind zones—westerlies, trade winds, polar easterlies, horse latitudes and the doldrums. Collectively these are considered to be the Earth's "prevailing" winds since their locations above the earth's surface, and the wind speeds and directions with which they are associated, are more or less predictable westerlies, also called "the prevailing westerlies". The westerlies are located in the meteorologically active 30-60 degree midlatitudes, where weather is most active at the Earth's surface. The westerlies are at their strongest in the winter. The trade winds are also called simply "the trades". The trades are powerful prevailing winds that blow across the tropics in the Northern Hemisphere from east to west. They may also be referred to as the tropical easterlies.

Some believe there are only four wind zones, with the Trade Winds and Horse Latitudes indistinguishable one from the other. Still other simplifiers believe the light winds around the Inter-Tropical Convergence Zone and equator do not qualify as a wind zone and the correct number is three (Trades, Westerlies and Polar Easterlies).

The westerlies are a major reason that the weather in the continental United States moves predominantly from west to east. The westerlies are located between the poles and the tropics and blow from west to east in the Northern Hemisphere, steering low and high pressure systems across the globe. By far the most common usage of the term "jet stream" focuses on the quasi-horizontal core of maximum winds found in mid-latitude westerlies, typically found in the upper/upper troposphere. <u>Westerlies are responsible for the weather in much of the world, including large portions of the United States.</u>

The trades, together with the westerlies, provided quick and reliable wind-enabled roundtrip routes for sailing ships crossing the Atlantic and Pacific Oceans from America to and from Europe and the Far East. Columbus knew this and used it to his advantage on his round trips. Columbus used the westerlies to propel him across the Atlantic and back to Europe. Tropical storms, including hurricanes, are formed within, and follow the paths of the trades. The trades also pick up and steer dust from the Sahara Desert across the South Atlantic into the Caribbean Sea and sometimes to the U.S. Knowing the wind speed and direction, temperature, humidity and barometric pressure at higher altitudes of the atmosphere is very important for weather forecasting. Radiosondes attached to weather balloons are launched twice a day from each of the 122 NWS Local Weather Forecasting Offices.

Measuring Wind Speed

Centuries ago, sailors had great difficulty measuring how fast their ships were travelling. At some point, an enterprising lad devised a system for doing so. The mainstay of the process was a very long piece of rope, with knots tied at consistent intervals along its length, and a piece of wood tied at one end to provide some stability when the contraption was dropped into the water. As the ship was sailing along the rope was dropped into the ocean, "wood end" first. The number of knots that slipped off the ship into the ocean from the ship were counted every 30 seconds. The ship's speed could then be calculated. As it turns out the distance between each knot was one nautical mile, so a knot is one nautical mile per hour. <u>As a general rule meteorologists express winds over land in miles per hour and winds over the ocean as knots</u>

But when informing the American general public about winds currently being experienced, or forecast to do so, on land or at sea, miles per hour are generally used. Though commonplace to mariners, the public has virtually no understanding or calibration of the word "knot" as applied to wind speed.

<u>The anemometer has been the standard for windspeed measurement for years.</u> It can easily handle all kins of winds and over quite a range, from light breezes to potentially catastrophic winds associated with tornadoes and hurricanes.

A simple algorithm converts the number of revolutions by the anemometer to wind speed. Ultrasonic sensors, or simply "sonics", are being used increasingly when highly accurate wind speed information is required, such as for detailed research climate data. On a typical sonic anemometer, a transducer sends a pulse of ultrasound from, for example, the north-facing side of the sensor; a microprocessor then measures the time it takes for that packet of air to make its way to the south-facing transducer. Without moving parts, the results are immediate and generally precise. On average over time, wind data gathered using cups and ultrasound are by and large the same. But sonic equipment does a better job handling gusts of wind and turbulence occurring over time spans of just a few seconds. This simply reflects the limitation of cup and paddle technology direction before it can begin collecting wind speed data.

The NWS found out last year that one of the buoys that used a commercial off the shelf sonic anemometer at a site in the middle of the Chesapeake Bay would occasionally read too high during gusty wind situations. Further testing in a wind tunnel at the NWS in Sterling determined it was producing faulty readings.

There are three primary and one secondary wind speed measurements made on a continuing basis at ground level; each has a specific set of criteria to ensure consistency and allow seamless comparisons with other wind events. These are 1) Primary – Wind Speed, Sustained Wind Speed, Wind

Gusts, and 2) Secondary – Squalls. From a wind measurement perspective "Ground Level" is defined as 10 meters (33 feet) above grade. This requirement makes it difficult for the amateur meteorologist to obtain what would be considered sufficiently accurate wind data at his or her home.

The Beaufort Scale

In 1805 Sir Francis Beaufort established an improved system for describing the force of winds while he was with the British Royal Navy. That scale still bears his name. It provides an <u>empirical description of wind speed based on observed conditions at sea</u>. In an age where weather instruments were not available, terribly inaccurate and/or very expensive, Beaufort's scale helped sailors estimate wind speed without needing to have such equipment. A pleasing aspect of the scale is the use of so-called General Terms to describe winds—light air, gentle breeze, moderate gale. The Beaufort Scale went through some changes in the 1940s when new categories were added to address higher winds associated with tropical cyclones. The Beaufort Scale has 12 levels of wind force. The first category is 0, virtually windless conditions so calm that the sea is "like a mirror" Category 12 is intended for hurricane force winds, though it has rarely been used since introduction of the Saffir-Simpson Scale for tropical cyclones in 1973.

Wind Speed Nomenclature

- "Wind Speed" is the average speed of the most recent two-minute period.
- "Sustained Wind Speed" is a 10 minute average.
    - o For tropical storms and hurricanes, a one-minute average is also provided for sustained wind speed.
- A "Wind Gust" is a sudden, brief increase in wind speed, typically lasting less than 20 seconds American Meteorological Society – (NWS is five seconds).
- A "Peak Wind Gust" is the highest instantaneous wind speed recorded during a specified period, typically the 24 hours from midnight to midnight with certain criteria being met.

- For specific circumstances, such as an active hurricane, derecho or severe thunderstorm, such readings are typically provided more frequently.
- A "Squall" is a strong wind characterized by a sudden onset.
    o It must maintain a windspeed of 16 knots for a minimum of two minutes.
    o This it from a wind gust.
- "Wind Shift" is a change in the direction of the wind of 45 degrees or more in less than 15 minutes with a sustained wind speed of 12 mph or more throughout the episode.
- By the way, with all these definitions being bandied about, the National Weather Service considers it to be "Windy" when winds reach 20-30 mph. (Why? No idea!)

Wind Rose

A "Wind Rose" is a chart that characterize the speed, strength and direction of winds at a specified location. The length of each "spoke" around the circle indicates the amount of time that the wind blows from a particular direction. Colors along the spokes indicate categories of wind speed. Such charts are created by gathering voluminous data, usually minute-by-minute.

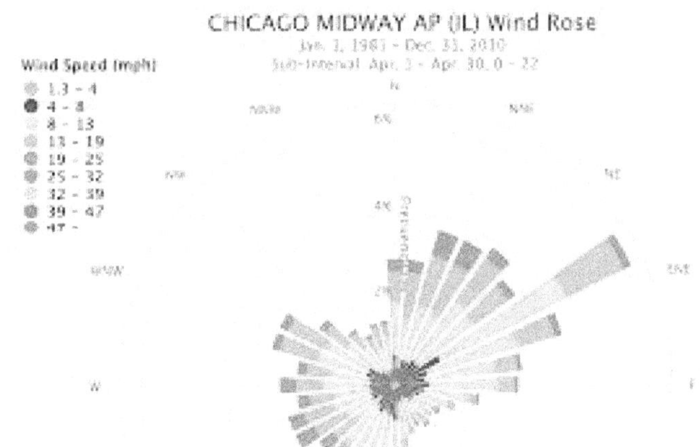

Here is an example of a Wind Rose.

Such charts are indispensable for a number of reasons. Architects use this data for sitting and designing buildings. Sailors use the data for establishing optimal routes between ports. Wind roses provide essential data for the owners of wind farms. The results allow them to site individual or multiple windmills in locations that provide the most wind for the greatest time, thereby maximizing the energy obtained from any single installation. It is now known that, had wind roses been used for the siting of the infamously cold and windy Candlestick Park in San Francisco, the stadium could have better been positioned just a few hundred yards north, where it would have been protected from the bitter prevailing westerly winds that often buffeted Candlestick.

Wind & Nature

For something as simple as moving air, wind plays a major role in Nature. Wind shapes, transports, scours, erodes, carries, disperses, brings together and otherwise affects every living thing and geological feature on the planet. And it is happening, this wind, all around us, all of the time—some days quiet, slow, almost imperceptible, then suddenly a storm and the wind's demeanor changes from friendly to threatening, from soft and gentle to foreboding and frightening. The roles wind plays in nature are not only fascinating, but they are also essential. Although many of the following "wind does this" topics (and others not included) could lend themselves to multi-page articles of their own, my intent is rather to introduce just a dozen or so of wind's handiworks, covering each sufficiently so that both their individual roles and the collective impact wind has on nature can be understood.

So, in no particular order, here are some examples of the wind's handiwork and roles.

Architect of Natural Beauty

Many forces are at work shaping the Earth's surface. Airborne sand and other solids contribute to the process of selective decay and erosion responsible for many of our most cherished sites of natural beauty. But in many cases forces other than wind predominate. For instance, freeze-thaw cycles are the primary sculptures of the hoodoos of Bryce Canyon National Park, and water-borne solids have acted likewise at Arches National Park. But there is little doubt that wind is king when it comes to shaping, moving and replenishing sand dunes.

Take Great Sand Dunes National Park as an example. Aside from its sheer size and beauty, Great Sand Dunes became a National Park because of a strong and dependable annual cycle of sand depletion and replenishment that works to keep the dunes generally stable. Every spring a portion of sand from the dunes is carried by two mountain streams to a nearby valley floor. As temperatures rise and summer approaches, the creeks dry up, and the exposed sand is blown by the wind back onto the dune field. The combination of seasonal wind speeds and directions, the supply of sand on the valley floor, and the sand recycling action of the streams and wind all contribute to making these sand dunes the tallest in America, 750 feet. The sand dunes at Indiana Dunes National Lakeshore on the shores of Lake Michigan have been on the move for years, gradually migrating inland and swallowing up a part of the parking lot in the process. It's projected that continued movement of the dunes will eventually affect nearby roads, one a state highway; time will tell.

Creator of Waves

Wind plays a major role in the development and intensification of ocean, sea and lake waves. Waves are generated by the sustained influence of the wind on the water's surface. Depending on the strength (force) of the wind and the time available to sustain that force on the water's surface, the resulting waves can range from a ripple to over 50 feet high.

The four primary factors that influence the likelihood and ultimate intensity of wave formation are, wind speed, uninterrupted distance of open water (fetch), wind duration and water depth. Waves provide a wide range of concerns and benefits. Waves magnify the effect of storm surge from landfalling hurricanes, limit sea travel and cause sea sickness. But larger waves also create a higher level of wave power (that is, the energy resulting and derived from moving water) thereby increasing the available energy for power generation. Within limits, larger waves enhance the surfing experience

Propeller of Birds.

Birds travel extraordinary distances during their annual migrations to winter nesting grounds. Some species, like the stork, complete a roundtrip that exceeds 4,000 miles. How can they possibly do this? After all it takes energy to propel anything (natural or man-made) through mile after mile of the atmosphere.

Answer? They let the wind help them. They let thermals do the heavy lifting. Thermals are areas of rising air (updrafts) caused by differential heating of the earth's surface. The storks literally "catch" a thermal, and circle higher and higher until it dissipates. They then begin a downward glide which has the maximum geometric horizontal component to take them as far as possible toward their ultimate destination. Another thermal is found, and the "thermal-glide" process continues. Although there are of course limits to what can be achieved—thermals need to be present (though orographic winds associated with mountain slopes will also do) and the process does not work at night. *But "thermal winds" certainly make the stork's job a lot easier.*

Accomplice of Allergies

Millions of people are allergic to all sorts of things, everything from cats to peanuts to shellfish to milk. Ragweed pollen is an especially notorious allergen. A single ragweed plant can produce a billion pollen grains. In one allergy season as much as a million tons of yellow ragweed pollen grains can blanket the United States. Pollen, when airborne, is per se invisible; it is also relatively heavy, compared to say bacteria. So, in a quiet, windless world, one could escape the ragweed allergy by staying away from the ragweed plants. Unfortunately, for most allergy sufferers, it is not a windless world. Wind scoops up the pollen grains and transports them for miles and miles. Some eventually reach your nose. Ah choo!

Determiner of Civilization

A visit to Mesa Verde National Park in southwestern Colorado provides an intriguing story of the Anasazi Indians who lived in cliff dwellings under, and grew corn, squash, maize and beans on the roofs (the Mesas) of their homes. The Anasazi civilization thrived during the 1100s and 1200s.

Successfully growing these crops required a minimum amount of rainfall and a minimum number of frost-free days. From the start the climate was close to these minimums. Over the course of decades, the climate in southwest Colorado became dryer and colder. A subtle change in wind patterns is suspected to have occurred over generations. Winds became more southerly; less moisture was carried by the winds; there were fewer rainy days. Where rain once fell dependably, rain no longer fell. The crops failed; the Anasazi moved on. The Mayan civilization and the Tang Dynasty also fell—maybe for the same reason. It didn't take much of a change in the prevailing winds—just a little less moisture.

Pollinator

Pollination occurs in plants when pollen grains from the male part of one flower (the anther) are transferred to the female part (the stigma/pistil) of another flower. Once pollination occurs, the fertilized flowers produce seeds, which enable the plant to reproduce and/or to form fruit, ensuring the survival of the species. There is certainly no shortage of pollen created by healthy plants—millions upon millions of the fine powdery, often yellow or green, grains are released by trees in a single shot

Countering such a seemingly massive availability of pollen is the exceedingly low probability that any single microscopic grain will actually reach and eventually stick itself to a stigma—perhaps only one in a million. Without a way to convey pollen from the immediate surroundings of the plant from which it was expelled, pollination would remain a very local event and individual plant species would "move out" into the world very slowly, indeed hardly at all. Insects and other animals do help to convey pollen, but *the real facilitator is the wind.* The great temperate forests, grasslands and savannahs of the world exist only because of wind-borne pollination. The sheer volume of pollen that can be carried, and the distances that need to be traversed, can only be accomplished by the wind. Plants also rely on the wind to disperse their seeds, a process called anemochory.

Replenisher of Soil Nutrients

Every situation in which wind transports dust is not detrimental; every scenario does not lead to a dust storm. In fact, there are a number of examples where wind's role in moving dust is essential for the well-being of the "receiving location". Like the Sahara/Sahel Conveyor Belt of northern Africa. <u>Wind is continually scouring the major deserts of Africa and sending its bounty of dust westward</u>. This dust contains all sorts of stuff—minerals, metals, spores, bacteria, pollen grains, dirt, silica, you name it. Zonal wind patterns change with the seasons; trade winds, the inter-tropical convergence zone, the breadth and speed of the prevailing westerlies. But the predominant flow is from Africa to the Americas, and an important, life-giving dust it is. Upwards of a billion tons of dust is carried each and every year by the winds coming off the west coast of northern Africa. It takes only 5-10 days for that dust to be carried by the trades to Brazil, and that dust contains essential nutrients to sustain life—calcium, iron, nitrogen and potassium. But the most important of these nutrients is phosphorus. Heavy rain quickly depletes the rainforest of this important nutrient. The annual transport of dust from Africa brings phosphorus to Brazilians soil and allows the forests to thrive. The eco-systems in Africa and South America are inextricably linked. It's the wind that does it.

Provider of Life-Sustaining Moisture

A monsoon is a seasonal change in the prevailing wind system of an area. It comes from the Arabic word "mason", meaning "season". A monsoon is the offshore/onshore breeze phenomenon on a mega-scale, monsoons are part of a year-long cycle of uneven heating and cooling of tropical and mid-latitude coastal regions. Monsoonal winds occur in many parts of the world, including the American desert Southwest, places like Arizona and New Mexico. In these regions the monsoon season runs from June 15 through September 30 and is characterized by a pronounced increase in dew point, thunderstorm activity and rainfall. By far the largest, most powerful and easily recognizable monsoonal winds are those of Asia, Africa and Australia, the so-called "monsoonal subcontinent" that includes Pakistan, Bangladesh and India.

Let's use India as an example of monsoonal subcontinent meteorological mechanics.

In the summer the land in northern India, for instance New Delhi, heats up much more quickly than the waters of the Indian Ocean. This hot continental air rises, causing an area of low pressure to form. Cooler moist air from the Indian Ocean is drawn into the void left by the rising hot air. Steady southwesterly winds carry the moisture-laden air into India's northern interior, where the moisture runs into and up the slopes of the Himalayan Mountains where it interacts with cooler air and condenses, and rain falls. The monsoon is vitally important to India, generally providing plentiful annual rain to fill aquifers and lakes, and to irrigate the soil.

Prodigious amounts of rain can fall from the monsoons. In Cherranpunji, India the average rainfall for the monsoonal month of January is 96 inches (8 feet!); in contrast, July averages less than an inch. By summer, winds have changed, and the process is reversed; the sea cools more slowly than the land; a low forms over the Indian Ocean and cool air is drawn in from the land. Dry winds blow from the northeast. When the monsoonal winds do not come, or come with less moisture than normal, a difficult year may await. But it was not just the farmers who eagerly awaited the arrival (and departure) of the monsoons.

Sailing the seas between Africa and India took advantage of the changing monsoonal winds, heading toward India as the summer monsoonal winds took shape and back to Madagascar as the winter north easterlies set up. Despite the generally welcome heavy rains brought on by the monsoons, there are years when monsoonal rains are heavy and relentless, and flooding occurs, sometimes accompanied by significant loss of life.

Destroyer of Plants

The wind can be a constant enemy of plant life, big and small. For instance, relentless winds can limit plant growth by inhibiting vertical expansion, as exemplified by the Divi Divi tree on Aruba (shown below).

High winds knock down, break and/or uproot trees via a process called "windthrow", and even if they survive such dramatic consequences, trees can have their branches and leaves ripped off, making them more susceptible to disease. Airborne dust damages plants through abrasion that erodes leaves, bark and seedlings.

Weapon of Warfare

Winds have played a major role in warfare throughout the ages. The Japanese attribute their continued existence to the wind from a series of typhoons that repelled Mongol fleets under the command of Kublai Khan from attacking their island nation in 1274 and 1281. The Japanese named the winds Kamikaze or "Divine Wind". Wind played a major role in deterring the Spanish Armada from invading England in 1588. That storm is known as the Protestant Wind. The storm was so powerful that remnants of the Armada's ships showed up on the shores of Ireland months later. On the other hand, favorable winds allowed William of Orange to invade England in 1688. Eventually the science of meteorology matured, and the element of surprise regarding specific wind events lessened. Reasons for events occurring no longer could be attributed to the gods. But even an improved understanding of wind and weather could not keep unexpected wind-related events from occurring. The so-called Egyptian khamsin wind, hot and dust-laden, seriously impaired French soldiers during Napoleon's North African campaign.

The same wind churned up massive sandstorms that affected both Allied and Axis forces and equipment during the Egyptian campaign of WWII, eventually bringing the battle to a standstill.

But WWII did eventually prove that there was a place for meteorology in war-game planning. Wind (and cloud cover) were the primary factors for deciding whether or not, and if so when, the Allies should attempt a marine landing in France. High winds would create high waves that would slow the overall operation, make beach assaults more difficult, and create seasickness D-Day was postponed from June 4 to June 6, 1944, based on weather reports that showed reduced cloud cover and gentler wind conditions on June 6. Interestingly, the Germans, also with rudimentary weather forecasting capabilities, forecast the same winds, but concluded that no attacks would occur during the entire period from June 3-7.

Contributor to Topsoil Creation

Soil (along with water of course) is the lifeblood of agriculture. The American Midwest and Plains states have an abundance of wonderfully rich topsoil. Surprisingly, wind has played a role in the development of this rich topsoil, and continues to do so. It takes a long time to create just an inch of topsoil, a numbingly slow process by which, at least initially, rock is broken down by a series of chemical and physical weathering processes. As it turns out the glaciers took care of a lot of the grinding, rubbing, compression, and freezing and thawing that oh so slowly reduced rock to dust. It was then up to the wind to transport this topsoil precursor to the American Midwest, a chore it did very well over the thousands of years following the last ice age. By the way there is a striking similarity between the airborne conveyance of essential nutrients like phosphorus from Africa to Brazil, and the creation of America's rich dirt from winds moving topsoil precursors from the remains of retreating glaciers. And today, throughout the world, including in America's breadbasket, a complex set of natural processes continues non-stop to enhance the quantity and quality of soil. It is fungi that take center stage in this process, of course aided by the wind.

Fungi orchestrate the relentless natural processes that break down dead organisms—leaves, twigs, flowers, animal flesh, excrement, paper, spiders, fish scales, bones. Fungi are everywhere and in enormous quantities—ten to fifteen thousand in a cubic yard of air, tasked only with returning spent animal and plant refuse to the earth and, in so doing, to in fact create new soil, albeit at a pace so slow it is imperceptible. But, cumulatively, over time, it happens. But fungi need to get out and about to do their job well. They need to travel to keep the transformation going, to get the job done. *To help? The wind of course.* Moving air ensures that fungi are replenished and mobile, able to maximize the output of their labor. Fungi spores are light and easily transported. To help this process work better many species have evolved over the eons to include a plant structure that allows fungi to literally be shot or expelled into the wind, further improving the ability of the wind to pick up and transport fungi spores over long distances.

Conveyor of Illness and Disease

The wind does not discriminate when it comes to deciding what it will, and what it won't, carry in its midst. Some things it transports are positive, like the moisture-laden monsoonal winds carried to India. Some, like pollutants, less so. Ironically enough, the immediate area creating smoke, ash, chemicals and other dangerous air borne. Pollutants often does not themselves experience the noxious compounds they produce. The wind does a thorough and effective job of picking up the bad stuff and transporting it tens or hundreds of miles before "dropping it". In the 1950s and 1960s, prevailing westerly winds over the continental United States caused sulfur dioxide, caustic smoke, and other pollutants from steelmaking operations to affect people on the East Coast more than their neighbors in Pittsburgh. Ingestion of pesticides and industrial pollutants originating on the steps of Mongolia, picked up by the winds from around the Aral Sea, and transported across the Pacific Ocean led to disease and poisoned breast milk among native Inuit in northern Alaska and Canada. Alkylation is a chemical process used in the oil refining industry in which light gaseous hydrocarbons are combined to produce the high octane components of gasoline.

Unfortunately, one of the best alkylates is hydrofluoric acid (HF), a chemical that can cause serious burns if touched and death if ingested. Refiners go to great lengths to protect their workers and the general public from harm should a leak occur in an Alkylation Unit. The refinery operators and first responders tasked with minimizing accidental human contact with gaseous HF acid need to know two important parameters before anything else—which way and how fast is the wind blowing. Since the generation of wind itself is enhanced by daytime heating of the air, nighttime winds (everything else remaining equal; that is, there are no frontal movements or thunderstorms) tend to be lighter, making it marginally easier to knock down and neutralize any escaping HF. This wind (or you might say this calm) phenomenon is known as Nighttime Stability.

Accelerator of Fire

We all know the fire triangle—a fuel source (paper, wood), air (or oxygen) and heat. For a fire to occur, all three ingredients must be present. Fires move with the wind, propelled, pushed forward. As the wind increases, two important factors come into play. First, the fire-front moves more quickly, and second, additional oxygen is added to the fire, increasing the fire's intensity. In Southern California, Santa Ana winds result from circulation of the atmosphere around a high pressure system east of the mountains. These hot, dry, high velocity winds move rapidly toward the coast, passing through canyons in the hills east of the Santa Monica Mountains that further intensify the wind's speed.

The Wind — Man, History & the Future

Sailing requires wind to create a force that acts on sails or similar surfaces to propel a craft on the surface of water. Over centuries man has sought the best materials for the sails themselves and the best way to arrange them in order to get the maximum benefit from the wind. Until the middle of the 19th century sailing ships were the primary means for marine commerce. This period was known as the "Age of Sails". A painted representation of a ship under sail was found in present day Kuwait; it suggests that man was using sailboats as far back as 5000-5500 B.C. Sailing provided superior means for trade, transport and warfare. Advances in sailing technology in the Middle Ages enabled Arab, Chinese, European and Indian explorers to make longer voyages and better survive stormy seas. Polynesians travelled vast distances throughout the South Pacific Ocean in search of habitable land. Eventually early man came to realize that wind energy, captured and harnessed, could provide the power to make many work tasks faster and easier. The benefits of wind energy then were the same as they are today: Wind is free, readily available and if you are in the right spot, reliable and predictable. Among the earliest records of man using wind to accomplish mechanical tasks occurred on Sri Lanka where monsoonal winds were used to power furnaces as early as 300 B.C. The monsoonal winds allowed the Sri Lankans to heat furnaces to as high as 2190 degrees Fahrenheit, hot enough to accomplish rudimentary metallurgical tasks. Heron of Alexandria developed a wind-powered organ in the 1st century A.D. This is generally regarded as the first machine powered by a windmill.

The first real use of wind energy to provide power to accomplish specific mechanical tasks took place in Persia in the 7th century A.D. Windmills having six to twelve rectangular cloth sails (helped along perhaps by additional sails with reed mating) were used for pumping water, grist milling and in the sugarcane industry. By the 10th and 11th centuries, the proliferation of wind-powered equipment being used in the Middle East caught the attention of European merchants, engineers and crusaders. Windmills began to appear in Europe in the 12th century. In the 14th century, Dutch windmills were being used to drain water from lakes and marshes of the Rhine River delta, a skill that matured and is still relevant in Holland today. Wind-driven technology continued to evolve. American colonists used windmills to irrigate crops, to grind wheat and corn, to pump water and to cut wood. By the 18th century, windmills were providing energy to pump water for salt-making operations on Bermuda and Cape Cod.

The first wind turbine used to directly produce electricity was built in Scotland in 1887 by Professor James Blyth, who used the energy to charge batteries (technically "accumulators") which in turn provided energy to light his cottage. Thus, his little cottage became the first house in the world to have its electricity provided by wind power. At about the same time in the United States, Charles Bush designed and built a monster wind turbine in Cleveland, Ohio. The turbine (shown below) weighed 4 tons, had 144 blades and rotors 56 feet in diameter, and powered a 12 kW generator.

The American Midwest was a very remote, rural, agrarian society before and up until the middle of the twentieth century. Farms in the Midwest, the Plains states and other parts of the United States were separated by many miles. It was difficult to imagine widespread distribution of electricity in this sort of setting. Small generators to provide energy to individual farms were at the time the only way to provide energy to America's widely dispersed locations. In the late 1920s, two brothers from Minnesota opened a factory to produce wind turbine generators for farm use, typically to provide sufficient power for lighting and for charging batteries that could be called upon later to power for certain appliances or motors. In 30 years, the brothers produced about 30,000 small wind turbines. By the 1930s, windmills were widely used to generate electricity on farms in the United States

In the meantime, work was underway to design and build a utility scale turbine whose output would be large enough to serve more than just one or a few farmhouses or residences. A forerunner of the modern utility scale wind generator was put into service near Yalta, on the Crimean Peninsula in 1931. This prototype, which ran until 1942, generated upwards of 100 kW of power. The world's first megawatt-size wind turbine connected to a local electrical distribution system began operating in 1941 on a mountaintop in Castleton, Vermont. It operated for 1100 hours before a blade failed. Wartime rationing of metals created shortages of spare parts and doomed further operation of the facility.

As the 1940s moved into the 1950s, distribution of electricity became commonplace in densely populated regions of America.

- However, the vast majority of this electricity was not generated by wind turbines.
- It was generated by burning fossil fuels like coal and fuel oil, with some contributions from hydroelectric sources.
- Wind was at best a minor contributor.
- This is still true today.

Some of what I came across while researching "wind" was really quite interesting and fun. And despite my misgivings at the beginning of the project that there would not be enough substantive information to populate multiple articles on wind, there really was a lot to write about. Once I embarked on this rather quirky subject (after all, who would pick wind among all the exciting meteorological topics out there?) I found more and more features of wind—it seemed to never stop. There was an array of topics and detail I had never even thought of—it was amazing.

In the process of studying the *wind* we learned that it has many capabilities. It can, among others

Carry	Collect	Erode	Destroy
Distribute	Intensify	Lift	Mix
Propel	Scatter	Sculpt	Shape

And there are lots of adjectives used to describe the *wind*, including:

Biting	Brisk	Blustery	Cutting
Dangerous	Fresh	Gentle	Light
Penetrating	Piercing	Powerful	Refreshing
Restless	Squally	Stiff	Stormy
Wandering	Wayward	Welcome	Wild

Wind has made its way into a number of <u>*colloquialisms, sayings and jargon.*</u>

- The calm before the storm
- Any way the wind blows
- To get wind of
- In the teeth of a gale
- To get one's second wind
- To get the wind knocked out of you
- To shoot the breeze
- To take the wind out of one's sails
- Three sheets to the wind
- To throw caution to the wind
- Wind at your back
- To be winded
- Wind bag, and perhaps quite appropriately as a last one:
- You don't need to be a weatherman to know which way the wind blows

<u>And where would sports be without the wind?</u>

- "The wind is blowing out tonight at Citi Field; I expect we will see a lot of home runs."
- "The Eagles have won the coin toss and deferred; they will receive the second half kick-off and will have the wind at their backs in the all-important 4th quarter."
- "It's a world record time in the women's downhill"; but not so quick. It now appears that the wind speed (literally the wind at her back helping propel her) was greater than allowed.

Windy Cities, Windy Countries; Calm Places; Exceptional Winds

The windiest place on Earth is Antarctica.

N.B. For this and all subsequent location-specific "annual average wind speeds" the speed is determined simply by averaging hourly readings occurring throughout the year and taken by approved instruments and, at NWS

Local Weather Forecasting Offices and other approved locations, at 10 meters (33 feet) above the surface (among other requirements)

The windiest city in the world is Wellington, New Zealand.

Among the countries with very low annual wind speeds are The Congo, Indonesia and Malaysia.

The highest wind speed ever recorded on the surface of the earth, and validated by the World Meteorological Organization, was a wind gust of 253 mph recorded near Barrow Island, Australia in conjunction with Typhoon Olivia in April 1996. That record eclipsed the previous record of 231 mph measured on Mount Washington, New Hampshire in 1934. The Mount Washington 231 mph wind speed remains the highest ever recorded that was not associated with a tropical storm (hurricane or typhoon). In March 2007 Typhoon Debbie made landfall near Arlie Beach, Australia with a 10-minute sustained wind of 119 mph, another record.

Windy and Not So Windy U.S. States and Cities

The windiest widespread portion of the continental U.S. extends from the Texas Panhandle north through much of Oklahoma, Kansas, Nebraska and South Dakota and into Minnesota.

- Of course there are local areas subjected to particular meteorological conditions that give rise to higher annual winds.

Rank	Windiest States	Calmest States
1	Nebraska	Mississippi
2	Kansas	Florida
3	South Dakota	Kentucky
4	North Dakota	Georgia
5	Iowa	Alabama

Below are five of the *windiest major cities* and five of the *calmest major cities* in the United States and their respective annual wind speeds.

Rank	Windiest Cities	Calmest Cities
1	Cold Bay, AK – 16.2 mph	Tucson, AZ – 1.1 mph
2	Dedham, MA – 15.0 mph	Eugene, OR – 1.5 mph
3	Dodge City, KS – 13.2 mph	Seattle, WA – 1.6 mph
4	Amarillo, TX – 12.8 mph	San Antonio, TX – 1.8 mph
5	Cheyenne, WY – 12.4 mph	Austin, TX – 2.1 mph

The Chicago Wind Lore

Many of you, I am sure, have heard Chicago referred to as *The Windy City* (even on the Weather Channel no less!). In fact, this moniker has nothing to do with atmospheric air movement. Rather it is said to describe the verbosity and boastfulness of Chicago's politicians. Whether you believe that or not, or have perhaps experienced the strong winds funneling through Chicago's skyscraper canyons and thought "Windy City" was an appropriate description, in fact the annual average wind speed in Chicago is lower than the average annual wind speed in, among others, Boston, Buffalo, Cleveland, Dallas, Kansas City, and Minneapolis.

Chicago's average annual wind speed is 10.3 mph.

But Boston is 12.3, Cleveland 10.5, and Milwaukee 11.5.

And if that wasn't enough, we find references to the wind everywhere we look or listen—song titles, adjectives, jargon, sports.

Wind is happening all around us, all of the time—sometimes quiet, slow, almost imperceptible, then suddenly pushing its energy toward us, strong and fearsome, unrelenting, imperiling us, our lives and our homes and workplaces. Then the next day, quiet again, but not for too long.

The specific roles wind plays in nature are both fascinating and essential.

Wind shapes, transports, scours, erodes, brings together and otherwise affects every living thing and geological feature on the planet.

The Earth is what it is today because of the wind.

With the things the wind can do the question remains: <u>Is there any other natural or manmade phenomenon that has the capabilities of the wind?</u>

Bob Dylan summed it up best:

"The answer, my friend, is blow' in in the wind

The answer is blow' in in the wind."

CHAPTER 4 — STORMS

A storm is a violent disturbance of the atmosphere accompanied by high winds and rain, thunder, lightning or snow. A storm is considered to be severe if it does or has the potential to cause damage, serious social disruption and/or loss of human life. Storms, including those of the severe variety, are an expected part of life In the United States. There are many different types of severe storms, and it is not unusual on any given day to have one or more areas of severe weather affecting many square miles of U.S. territory. The National Weather Service regularly issues watches and warnings about such unsettled weather. Where one lives defines the types of weather one is likely to experience. Each type of storm has unique characteristics and each severe storm has one or more inherent vulnerabilities/risks accompanying it. But no weather events are more destructive, life-threatening or downright frightening than hurricanes and tornadoes.

The sheer power of these storms, and the energy embedded within them, is almost unimaginable. Both can, and do, cause tremendous damage to physical structures like houses, schools and churches; to trees, crops and other vegetation; and to infrastructure like roads, bridges, ports and electric power supplies. Although there are distinct differences between these two meteorological monsters, there is <u>one common denominator—wind!</u>

This article will describe how hurricanes and tornadoes get started and how the destructive winds that accompany them are formed and sustained, and it will review some of specially exceptional hurricanes and tornadoes that have affected mankind in general and the United States in particular.

Hurricanes

In terms of the power and damage inflicted on each square mile, multiplied by the number of such square miles so affected, <u>hurricanes are without question the most violent storms on Earth</u>. Whatever they are called, tropical cyclones all form the same way. The process of hurricane development is complex, requiring tropical instability, sufficiently warm water and air temperatures, high humidity, low barometric pressure and the absence of high altitude winds. Favorable development and strengthening occur when ocean water is 81 degrees Fahrenheit or warmer. Many Atlantic Basin hurricanes have as their origins tropical imbalances called "waves" which leave the western coast of Africa, heading westbound. Waves cause warm, moist air from the ocean to rise and churn. The Coriolis Effect is the name of the physics principles that deflects moving objects to the right in the Northern hemisphere. The Coriolis Effect is an important feature for tracking hurricane development and movement. Below are the official National Weather Service names and criteria for tropical systems in ascending order of wind strength. When winds in the rotating low pressure system reach 39 mph, the system is called a tropical cyclone or a precursor to a tropical storm. Although not used frequently, and not an actual storm, the information it generates is still valuable from a forecaster's perspective.

- A tropical wave is a cluster of thunderstorms moving westward within a low pressure trough that has the potential to become a tropical depression.
- A tropical depression is a tropical cyclone with maximum sustained winds of less than 39 mph.
- A tropical storm is a system of thunderstorms characterized by surface circulation and accompanied by sustained winds of 39-73 mph.
- A hurricane has sustained winds of 74 mph or more; it is at this point that an eye may appear.
- Cyclones, typhoons and hurricanes are all tropical storms, categorized by where in the world they each occur.
- The word "hurricane" only applies to tropical cyclones in the North Atlantic, the Caribbean and the Eastern Pacific.
    - o In the Western Pacific the term used is typhoon.

o In the Indian Ocean it is cyclone.

Once the storm becomes a hurricane, it is assigned a name. The practice of naming first women then women and men goes back hundreds of years. Initially, hurricanes in the West Indies were named after a particular saint's day. Australian hurricanes were first given names in the 19th century by Clement Wragge, an Australian meteorologist. *Unfortunately Clement tended to assign names to people he didn't like.* Clearly this process wasn't going to last too long. A more formal process of using names for hurricanes began in 1935 under the auspices of the World Meteorological Organization (WMO). Under pressure from women's groups and others, men's names were added to the hurricane list in 1979. When an exceptionally destructive storm occurs, especially one causing significant loss of life, the World Meteorological Organization "retires" the name, meaning it will never be used again.

Tropical systems, regardless of their wind speed and strength, will generally continue to grow until impacted by meteorological features such as wind shear or cold water that interfere with the storm's rotation or its condensation engine. Storm winds play a major role in increasing the height and power of the storm surge as it moves across open water and then comes ashore. The surge is literally a wall of water that can reach 10 feet or more in height as it rises out of the sea and rushes ashore. The severity of the storm surge is influenced by the storm's intensity, the size of its wind field, and the geographic make-up at landfall.

As the storm continues to strengthen, a small eye develops around the center of the storm. Despite everything going on around it, the eye is an amazingly quiet place with virtually no wind and no rain. Just blue skies. The eye is a particularly unique aspect of hurricanes. Eye wall formation and sustenance depend on, and are fueled by, a remarkably efficient use of warm ocean water over which the storm is moving. As long as warm air and water temperatures are maintained and destructive forces such as high level shearing winds and/or land masses are not encountered, <u>the hurricane can become remarkably powerful</u>. Water temperatures are crucial to hurricane development and storm winds play a major role in increasing the height and power of the storm surge as it moves across open water and then comes ashore. Storm surge is a pronounced and abnormal rise in seawater level that accompanies landfalling hurricanes. Given the opportunity to continue to strengthen over open ocean waters hurricanes can grow into the <u>most powerful storms experienced anywhere on Earth.</u> This hurricane power includes the eyewall winds themselves, that is, the winds within the immediate vicinity of the eyewall, and the winds in the outer reaches of the storm's circulation, much diminished but still capable of significant damage and risks to buildings and people. EF5 tornadoes are just as strong as any Category Five hurricane, but cover a much smaller area.

Prodigious quantities (10-20" or more) rain can accompany hurricanes as they move across open ocean waters and come ashore, but occasionally a fast moving one might only drop 1-3 inches of rain.

Very little on Earth can withstand the raw power of storm surges, particularly humans, who are enormously vulnerable. It was quite instructive to see a two-story, cinder block apartment house our son lived in while with the Navy literally wiped clean by the storm surge of Hurricane Katrina. Not a cinder block to be found. Unimaginable power. <u>Storm surge is a process that is *responsible for about half of all hurricane-related deaths*</u>, and for much of the structural damage that occurs from hurricanes.

Hurricane Winds

There is increasing evidence that small rotational features (mini tornadoes) called "mesovortices", which are embedded in the eye walls of tropical cyclones, contribute to the overall power of tropical cyclones by increasing the cumulative wind speed of the entire storm system. This phenomenon also helps explain why damage seen in certain hurricanes is seemingly greater than could be explained by the rotational speed of the eyewall alone. Hurricanes can range in size from 100-450 miles in diameter; the eye of a hurricane can have a diameter of up to 30 miles. In April 1996 Tropical Cyclone Olivia struck Barrow Island Australia.

A mesovortex within Olivia's eyewall produced a wind gust measured at 253 mph. Four years later, the World Meteorological Organization validated the data, making that reading the highest wind speed ever recorded on the Earth's surface. Prior to that Australian wind event, the highest surface wind velocity established in 1934 on 6,288 foot high Mount Washington, NH. But of course the winds on Mount Washington received no help from tropical systems. Tropical Cyclone Debbie made landfall in March 2017 near Airlie Beach Queensland, Australia with a peak gust of 160 mph and a 10-minute sustained wind of 119 mph. Of the many hurricanes that have occurred in the Atlantic Basin in recent decades, Hurricane Ethel (September 1960) was one of the strongest, sporting a sustained wind speed of 160 mph as it wandered around the Gulf of Mexico. It should also be remembered that hurricanes also frequently spawn tornadoes when they make landfall. The "Great Hurricane of 1780" with winds estimated at 200 mph is thought to be the deadliest in recorded history, with over 20,000 estimated dead as it traveled over eight Caribbean islands and Bermuda.

Since 1851 just short of 300 Category 3+ hurricanes have made landfall somewhere along the approximately 3100 miles of U.S. coastline between Brownsville, TX and Eastport, ME. That's only about two landfalls a year, probably a lot less than most people think. Almost 70% of U.S. landfalling hurricanes had peak winds less than 115 mph. A major hurricane (which has what are considered to be catastrophic winds of greater than 115 mph) strikes the U.S. East Coast on average about once every other year. Storms with such winds have led to massive damage and loss of life. Unfortunately, many ane-mometers themselves are not capable of withstanding the winds for measuring wind speed generated by the which they are supposed to be measuring.

The most intense U.S. hurricane, as measured by barometric pressure came ashore on the Florida Keys on Labor Day 1935, with an estimated maximum wind gust of 200 mph. A similar velocity was estimated as the maximum wind speed for Hurricane Camille, which came ashore in Missis-sippi in 1969, destroying every wind measuring instrument in its path. Then hurricane Andrew struck South Florida in 1992, causing widespread damage. A wind gust of 177 mph was measured at a private residence, though never validated by the National Weather Service.

The Saffir-Simpson Scale was developed in 1971 to help characterize threats from hurricanes and to have a yardstick to compare damage to build-ings, vehicles and infrastructure based on wind speed. Its creators were Her-bert Saffir, a civil engineer in private practice, and Robert Simpson who was at the time the Director of the National Hurricane Center (NHC). The scale was introduced to the general public in 1973. It became an accepted bit of hurricane meteorology within just a few years. Hurricanes are categorized based on the strength of the storm's maximum sustained winds. Hurricanes are separated into five levels of increasing intensity and associated damage. Although the Saffir-Simpson Scale has proved a useful tool, meteorologists, emergency management personnel and engineers have begun to understand that storm surge and inland flooding following the hurricane's transition to a land-based storm are responsible for far more deaths than from the wind. Forecasts increasingly provide warnings to that end.

The Tornado

The tornado is a violently rotating column of air that extends from the Earth's surface up and into the lower reaches the atmosphere. It is this spin that has led the general public to affectionately be called "twisters." The air in the center of a tornado is actually rising; in this regard, tornadoes can be thought of as "sucking" up air/debris from the surface into the upper parts of the tornado-bearing storm. To be classified as a tornado, rotation must exist both inside the thunderstorm and at the Earth's surface. Every thunderstorm is created by an updraft, and every tornado has a downdraft in which precipitation falls from the storm.

Updrafts and downdrafts are upward-moving and downward-moving air currents, respectively, caused by changes in localized absorption of heat from the Sun, which in turn causes some of the warmer air to rise, replaced in turn by descending cooler air. Updrafts and downdrafts also occur as part of the turbulence that is created when air passes over topographic barriers such as mountains.

Strong updrafts and downdrafts occur in thunderstorms as well. Updrafts characterize a storm's early development, during which warm air rises to the level where condensation begins, and precipitation starts. In a mature storm, updrafts are present alongside downdrafts caused by falling precipitation. These downdrafts, originating at high levels, contain cold, dense air that spreads out to the ground as cold air. The sharp changes in wind direction associated with downdrafts near the ground are a threat to aircraft during landing and takeoff.

It is fair to say that the intricacies of tornado development are still not completely understood by the meteorological community. But much is known. The most intense tornadoes are associated with, and emerge from, strong supercell thunderstorms. Supercell thunderstorms are characterized by the presence of a deep, persistent rotating updraft, technically a meso-cyclone. For tornados to form and grow, both the right ingredients and the right air movements are essential. Warm moist air near the earth's surface and relatively cool, dry air above it set the stage for tornado development—the buoyant warm air rises and comes in contact with cool air downdrafts. It is at this point in the tornado development process that the winds take over. Wind shear, downdrafts and updrafts add all sorts of discombobulation in the atmosphere. Air moves in horizontal and vertical directions; warm air continues to feed the updraft; a downdraft is established and fed with cool, dry air; regions of unstable air produce areas of intense vertical growth.

If the warm air rises quickly enough, and the fledgling cyclone is simulta-neously subject to spin from any number of factors, such as cooler air rushing in to fill the void created by the rising warm air, a vortex may develop. Ultimately it is this spin that drives these storms and creates such potentially devastating energy. If the circulation continues, both at the surface and at higher elevations, a tornado may begin to take shape. Given the right combination of temperature, humidity, condensation, and wind at the surface and aloft, and other factors known and not known or fully understood, nature can create some of the stron-gest storms on the surface of the Earth—the winds generated by tornadoes. The footprint created by a given tornado can range from narrow to wide.

There are two regions of the world where most of the world's tornadoes occur. One comprises Bangladesh and northeastern India. The other is the North American Midwest. Both regions lie east of a major mountain range (in the U.S. The Rockies) and north of a warm ocean (in the U.S. the Gulf of Mexico). It is the combined influence of cool, dry air to the north with warm humid air to the south, unimpeded by any significant geographic features, that allows tornado formation to occur. Tornadoes are also experienced in other countries like New Zealand, Australia, Canada and Germany.

There are about a thousand tornadoes every year in the United States, most occurring between 4 p.m. and 9 p.m. in the spring and early summer in a region known as "Tornado Alley" that comprises Oklahoma, Kansas, northern Texas, eastern Colorado, Nebraska, eastern South Dakota and southern Minnesota. Tornadoes are the most likely to occur in the months of May and June in North America, when the meteorological conditions that create and sustain tornadoes are most prevalent and energetic.

Most tornadoes last less than 10 minutes, have wind speeds of less than 110 mph, are about 250 feet across and travel just a few miles. Although far fewer in number, large, potentially destructive and deadly tornadoes may have wind speeds exceeding 300 mph, be more than two miles wide and stay on the ground for more than 20 miles. Unlike hurricanes, whose eyes may have diameters of 15-20 miles or more, the energy in tornadoes is, and the resulting impacts are, concentrated along a relatively narrow path. About 60 people die annually in the United States from tornadoes, most from flying debris. Half of the deaths can be attributed to the strongest 1% of the storm.

The deadliest tornado disaster in United States history occurred on March 1, 1925, when a single monstrous storm wreaked havoc during a 200+ mile journey from Missouri to Indiana, destroying nine towns and killing 695 people. The storm was so wide and so destructive it lacked the typical funnel shape of most storms. One can only imagine the extent of human death and injury and property damage had this same storm taken this same path with the same energy 80 years later. The most extensive tornado outbreak in U.S. history occurred from April 3-4, 1974. Called the "Jumbo Outbreak", the event consisted of an incredible 148 individual tornadoes in 11 states. At one point as many as 20 individual tornadoes were on the ground at the same time.

Much as the Saffir-Simpson Scale was developed to provide a convenient way to compare hurricane intensities and infrastructure damage it causes, so was a similar process developed for tornados. The process allows one to rate tornado intensity based on the damage it caused. Ted Ed Fujita was asked to come to the University of Chicago in 1953 to continue his research on downdrafts there. The Fujita scale was first introduced in 1971, a joint effort between Fujita and Allen Parson, then head of the National Severe Storm Forecast Center. Though each damage level is associated with wind speed, the Fujita Scale is effectively a damage tool. An enhanced Fujita ("EF") Scale was developed in 2007 to more accurately match wind speeds to the severity of damage caused by tornadoes.

Thunderstorms

Thunderstorms are frequent visitors to the United States, and more times than not, an unwelcome guest. Thunderstorms are a common occurrence, especially in the U.S. South, where thunderstorms occur every month of the year. It is estimated that there are approximately 100,000 thunderstorms every year in the U.S., about 10% of which are considered severe, meaning the storm is capable of generating half-inch size hail and/or winds in excess of 58 mph. Thunderstorms typically form in the late afternoon and evening when temperatures are at their highest. Many thunderstorms create distinctive, anvil-shaped clouds that reach heights of 40,000 feet or more, often in apparently quiet isolation. With all that moist air reaching into the coldest regions of the atmosphere, the "evaporate-lift-condense-precipitate" cycle is working smoothly, assisted by the larger scale Ice Phase Process and possibly the Collision-Coalescence Process. Prodigious rainfall rates, lots of lightning and thunder, hail and high winds are to be expected.

Thunderstorms develop in one of two ways. The most common and recognizable are thunderstorms associated with frontal movements. And usually this is a cold front plowing into and displacing an area of typically warmer, more humid air. Depending on the speed of the front and the depth of the unsettled weather through which the front must pass, thunderstorms can be over in as little as a few minutes or linger for hours. Currently, anyone with an iPhone can find the location of the front approaching their location. With some simple mathematics can "calculate" the arrival time of the front. And, by knowing what colors are associated with what Doppler Radar impulses, one can make a reasonable estimate of overall storm intensity. The other type of thunderstorm is called the Pop-up Thunderstorm. These storms are not associated with frontal movements and can literally come out of nowhere. All that energy can explode with an ear-splitting crackle and a simultaneous thunderclap that can occur far too close to your neighborhood, your backyard or your house itself. Pop-up thunderstorms occur when strong local convective energy causes rapid ascent of hot, humid air into the coldest reaches of the Troposphere. Pop-up thunderstorms can retain their integrity and stability for miles as a still-dangerous renegade storm.

Air Masses and Weather Fronts

An air mass is a large body of air with generally uniform temperature and humidity. The motion of an air mass coincides with the air flow in the upper atmosphere. As the jet streams change their position and intensity, so too do the motion and strength of the air masses. When air masses converge, they form boundaries called "fronts".

A weather front defines the boundary between air masses of different temperatures and/or water vapor contents. When a front goes through your little spot on the earth's surface you are said to have just experienced a frontal passage. Weather maps show the location of fronts and the high and low-pressure systems with which they are aligned/attached. Each front and pressure system has its own color and shape. Fronts don't exist only on the earth's surface; they have a verticality that extends dozens of miles up into the troposphere. Air masses extend up many hundreds of miles from the earth's surface. Air masses, pressure systems and fronts are essential elements of weather forecasting. In this complicated world of meteorology, there are only four types of weather fronts and, of these, only two (cold and warm) are mentioned enough for the general public to appreciate what they are.

The cold front is the most common, recognizable and understandable of the weather fronts. Cold fronts displace relatively warmer, more humid air masses with relatively cooler, less humid air. To nobody's astonishment, the predominant direction for cold front movement in the continental U.S. is northwest to southeast. Especially in the summer, cold fronts are associated with severe weather and can spawn thunderstorms and tornados. A cold front has about twice the forward speed of a warm front. Cold fronts are represented on weather maps by solid blue lines and blue triangles that point in the approximate direction the front is moving. The Mid-Atlantic region of the U.S. may experience 30 or more cold front passages a year.

The so-called back door cold front is a unique frontal feature periodically affecting a small portion of the U.S., typically southeast New England. Unlike cold fronts moving across the Continental U.S. from west to east, back door cold fronts typically move from northeast to southwest (through the back door) and derive their "cold "from the cold water of the Canadian Maritimes. Back door cold fronts periodically affect Boston, other parts of New England, and points south and west, and can lead to dramatic drops in temperature in just a matter of hours, sometimes catching weather forecasters completely by surprise. As an example, assume a back door cold front is moving to the west, having first been spotted near Providence, RI. and is now situated near New Haven Connecticut. Ahead of the front in Rye, NY, folks are likely enjoying a wonderful early summer day with temperatures close to 80F. But behind the back door cold front in New London, CT, just 50 or so miles to the east, it's an entirely different story, just 55F and a lot more clouds as well.

The warm front marks the boundary between a warm air mass and the colder, drier air it is displacing. Warm fronts move more slowly than cold fronts because it is more difficult for the light warm air to actually push the cold, dense air away. The tendency is for the warm air to move up and over the top of the cold air mass. Warm fronts gradually displace the colder air in their paths. As the warm air is lifted up and into the troposphere, clouds and rain form, often well ahead of the actual front. Unlike cold fronts, warm front passages are not accompanied by heavy winds. On the U.S. weather map warm fronts are characterized by solid red lines and red semicircles.

Stationary fronts occur when two air masses collide but neither has the energy to displace the other. Opposing cold and warm fronts stop moving. Stationary fronts can stay stagnant for days. Eventually a meteorological event such as a strengthening high-pressure system or a change in the jet stream will nudge the stationary front into motion, in which case it becomes either a cold front or a warm front. On a weather map a stationary front has symbols of both cold and warm fronts, shown as alternating red semicircles and blue triangles.

Now it gets complicated. Few people have heard of an Occluded Front. In fact, the dynamics associated with occluded fronts are such that even skilled enthusiasts are hard-pressed to explain them in great detail. It is highly unlikely that your favorite TV weather personality is ever going to say the words while on the air. An Occluded Front is formed when a speedy cold front over-run a lollygagging warm front. These sudden discombobulations put into motion a relatively complicated set of atmospheric actions involving the original cold and warm fronts, a new front, the "occluded front", and a low-pressure system. The fronts are referred to as occluded because warm air becomes separated (or occluded) from the accompanying low-pressure system. Occluded fronts are unusual. However, when they do form, they are of interest to forecasters. As it turns out the physical properties that create and maintain occluded fronts are of more interest than the fronts themselves. On a map, occluded fronts are represented by a solid purple line with alternating purple triangles and circle pointing in the direction the front is moving.

Lightning

The meteorological feature we call "lightning" may be the most beautiful and awe-inspiring of all the weather phenomena we experience. Especially in the darkness of night, lighting up our surroundings with unimaginable brightness, though usually for just a few seconds. A little scary too; something to stay away from; capable of generating energy and temperatures not unlike those of the Sun.

In this article we'll take a look at how lightning is formed. We'll review some of the special and extraordinary characteristics and features of lightning. We'll talk about the dangers associated with lightning and how to minimize your risk of becoming a lightning strike victim. And we'll of course review lightning's companion—thunder.

How Does Lightning Form?

The bolt of lightning you see during a thunderstorm is the visible discharge of electrical energy that occurs when areas of strong positive and strong negative charges interact. Thunderstorm creation is a rapid, rambunctious, violent process. Ascending water droplets are propelled into colder reaches of the storm clouds by turbulent updrafts. The water droplets eventually freeze; crystals are formed. At the same time that updrafts are active, downdrafts are pushing particles into the lower reaches of the clouds. Physics says that when two objects are rubbed together, one loses electrons and becomes more positively charged, while the other "collects" the stripped-off electrons and assumes a more negative charge. Many small bits of ice bump into each other as they move helter-skelter inside the thunderhead. Untold millions of exchanges take place every second during thunderstorm development. Slowly these collisions fill the cloud with increasingly electrified and powerful electrical charges. Eventually the heavier negatively charged particles congregate at the bottom of the cloud. Lighter positive charges gather higher up inside the thunderhead. The higher the cloud top the greater the space for particle interaction; the more charges produced, and the greater the storm severity.

Charge Field Alignment and Engagements

Negative and positive charges have a strong natural affinity for each other. As the charge fields intensify inside the cloud, objects on the ground are affected by the strong negative charge at the bottom of the cloud. As a storm moves over ground, the strong negative charges in the cloud attract positive charges on the ground. Positive charges are attracted to, and move up the tallest objects like trees, chimneys and telephone poles. A so-called "stepped leader" of negative charges descends from the cloud, seeking the best path for charges to get to "ground". As the negative charges gets closer to the ground, feelers with positive charges, called among others "positive streamers", shoot up to meet the descending negative charges. Once joined, it takes just a matter of milliseconds before a bolt of lightning is created. The lightning strike is actually two strikes. The first bolt (the leader bolt) zig zags downward to complete the circuit between cloud and ground. This is followed a split second later by a huge surge (the return stroke) that shoots back up the path just created. The speed of the lightning bolt as it makes this round trip is approximately 270,000 mph. Lightning strikes carry enormous amounts of electrical energy and are by far the most destructive of any localized weather phenomenon.

Characteristics of Lightning

Lightning is almost always associated with thunderstorms. Lightning does occur occasionally during snow storms. About 20 million bolts of lightning are generated in the Earth's atmosphere every year. Two-thirds of all lightning is generated in the tropics. There are close to a thousand lightning strikes in a typical thunderstorm. The Empire State Building in New York City is struck by lightning about 100 times a year. A typical lightning bolt is surprisingly narrow, just half an inch in diameter. In a fraction of a second lightning heats the surrounding air to an unimaginable 54,000 degrees Fahrenheit, six times as hot as the surface of the Sun. The electrical discharge from a lightning bolt averages about 1.5 million volts. The light from a typical lightning bolt is roughly equivalent to the amount of illumination from 100 million light bulbs.

There are several types of lightning, the most common being intra-cloud lightning. Intra-cloud lightning occurs completely inside one or more clouds jumping between areas with different electrical charges. It is referred to as cloud-to-cloud lightning. It is also called sheet lightning, a recognition of its appearance as it lights up the sky. The type of lightning with which we are most familiar is that which accompanies thunderstorm activity. This is cloud-to-ground lightning. Its unique appearance is one of power and beauty, a large bolt crashing to the Earth. But, alas, this is just an optical illusion; the most powerful stroke of lightning is the one barreling skyward. There are 5-10 times as many instances of intra-cloud lightning as there are cloud to ground bolts. Thunderstorms accompanied by lightning are a common feature in the U.S. West during the summer months. It is estimated that about 10-15% of all U.S. forest fires can be attributed to lightning.

Lightning Safety

No place outside is considered a safe location to ride out a thunderstorm. Never take shelter under a tree or near other tall objects which act as collectors of positive electrical charges. This could put one right in the path of the lightning bolt. The chance of being struck by lightning is about 1 in 600,000. About a quarter of people who are hit and injured by lightning, about 300 a year, die as a result, primarily from heart attacks in which the strong electrical nature of the lightning stroke stops the heart. A person's odds of death from a lightning strike are similar to those of being a railroad passenger or from bee stings. A lightning bolt can strike the Earth miles from the actual thunderstorm. This is why the mere sound of thunder at a community pool miles from the actual storm will cause the pool to close for at least 30 minutes. The states in which risks are highest for being struck by lightning include Florida, Texas and Colorado. The lowest probability is in Alaska, Rhode Island and Oregon.

Lightning rods are metal pieces placed at the highest point of a house, above the roof line, designed to "catch" lightning strikes and channel the energy harmlessly "to Earth", usually into several feet below grade. Without lightning rods homes and other buildings were routinely being stuck by lightning and burned to the ground. Benjamin Franklin saw the solution and invented the first lightning rod around 1750. Many houses built until just recently had massive amounts of metal in their plumbing systems, adding more "lightning rods" to the structure. Nowadays more and more PVC is being used to build these same plumbing systems—PVC is lighter and easier to join and otherwise work with; but it's nonconductive and of no value at all as a "lightning rod". Lately, researchers have found a way to "neutralize" and otherwise interfere with the "mechanics" of lightning strikes. Early days and a long way to go before commercial products are available.

Lightning Causes Thunder.

Air heated by the creation of lightning expands rapidly. This expansion in turn creates a shockwave that compresses the surrounding air, which then cools and contracts. An initial "crack" of thunder may be followed by rumbles as the column of air continues to vibrate. Light travels through the air roughly a million times faster than sound does so you are aware of the lightning strike well before you hear the thunder. To calculate how far away a lightning bolt is: after you see the flash, count the number of seconds until you hear thunder. Divide the seconds you counted by five. The resulting number is the distance in miles you are from the lightning bolt. Lightning starts about 10-15% of all the forest fires. The large-scale meteorological features that lead to accumulating snow vary markedly from one part of the world to another. On the U.S. East Coast, from the Mid-Atlantic States north to New England, the Nor'easter is the predominant heavy snowmaker.

Nor' easters

The National Weather Service defines a Nor'easter as "a strong low pressure system that affects the Mid-Atlantic and New England States". It can form over land or over the coastal waters. These winter weather events are notorious for producing heavy snow, rain, and tremendous waves that crash onto Atlantic beaches, often causing beach erosion and structural damage. Wind gusts associated with these storms can exceed hurricane force in intensity. A nor'easter gets its name from the continuously strong northeasterly winds blowing in from the ocean ahead of the storm and over the coastal areas.

A nor'easter is a large extratropical low-pressure system. "Extratropical" means that nor'easters are active outside the tropics; hurricanes on the other hand are tropical in nature. A nor'easter usually intensifies, sometimes dramatically, as it moves northeastward along the coast, typically reaching peak intensity in offshore New England. Cold air that is drawn into its circulation provides the energy to sustain the strength of the storm. Most form between September and April. But nor'easters technically can, and do, form at any time of the year. The storms are called nor'easters because they track from southwest to northeast, often embedded in, and/or propelled by, the Gulf Stream. The winds that precede and accompany the storms, and which bring the harshest weather, including the heaviest snows, are generally northeasterly in direction.

Nor'easters produce heavy, wet snow in, for instance, the Pocono Mountains of eastern Pennsylvania, and in much of New England, including the White and Green Mountains of New Hampshire and Vermont respectively. The heaviest snow typically falls to the west of the storm's center. Nor'easters can produce prodigious amounts of snow, with 8"-12" common in many storms. It is not unusual for nor'easters to drop 18"-24" under highly favorable conditions. Occasionally more than 30" is recorded. The large-scale meteorological features that lead to accumulating snow vary markedly from one part of the world to another. On the U.S. East Coast, from the Mid-Atlantic states north to New England, the nor'easter is the predominant heavy snowmaker.

Over the years, warming temperatures have led to changes in annual snowfall totals and the severity of significant snowfall events. Denver, for example, has seen its annual snowfall slowly diminish, from close to 60" in the 5-year period 1975-1979 to 54" from 1995-1999 to 45" from 2015-2019. Conversely, areas near the Great Lakes affected by lake effect snow have seen more frequent, and more significant, lake effect events and correspondingly higher snow totals. This increased snow in lake effect snow susceptible areas is due to the fact that it is taking much longer for the lakes to freeze over. Major cities along the 95 Corridor of the U.S. East Coast have had a significant increase in major snowstorm frequency and severity in just the last 30 years.

Where	Number of Storms with Highest Snow Totals Since
Washington D.C.	• Seven of the top ten have occurred since 1979.
Philadelphia	• All five of the highest have occurred since 1953. • The top three have all occurred since 1996.
New York City	• Seven of the nine have occurred since 1983. • Three of the five have occurred in the last 10 years.
Boston	• Eight of the ten have occurred since 1978. • Five of the eight have occurred since 2003.

The number of nor'easters occurring along the Eastern Seaboard varies from year to year. As does the severity of the storms. Some years there may be as many as 10-15 nor'easters. In other years just a couple. A major nor'easter with heavy precipitation, punishing tides and beach erosion, and winds approaching hurricane force affects the Northeast about once every five years or so. Over the years, nor'easters have been responsible for billions of dollars of damage, and hundreds of deaths and injuries. Because the paths of so many nor'easters run parallel to the megalopolis that stretches along the "95 Corridor" from Boston to Washington D.C., the potential for any nor'easter to create significant human misery and housing/infrastructure damage is substantial.

It was a nor'easter that led Benjamin Franklin to correctly understand the movement of storms along the Eastern Seaboard one night in 1743. Mr. Franklin was denied the clear skies needed to witness a long lunar eclipse. Weeks later, after exchanging a series of letters, he learned that his brother in Boston had enjoyed clear skies the night of the eclipse, but had the next day suffered through a storm that sounded similar to the one that had afflicted folks in Philadelphia 24 hours earlier. That information convinced Benjamin Franklin that this and other weather features were moving up the coast in a northeasterly direction.

Nor'easter Life Cycle

Nor'easters develop because of the difference in temperature and dew point between warm, humid air being lifted northward by the Gulf Stream and cold, dry air moving south from Canada. When these air masses collide, heavy, cold, dry air sinks. The lighter, warmer, more humid air rises. The turbulent air begins to spin. A low-pressure system begins to take shape.

If there is a major difference in temperature and humidity between the colliding air masses, and other meteorological conditions are favorable, the low pressure system may strengthen quickly and develop into a Nor'easter. Such storms can also form farther south, in the Gulf of Mexico, and occasionally as far north as New England. They can also have, as their origins, low pressure systems tracking across the continental United States.

The confluence of a number of geological and oceanographic features, and favorable meteorological conditions, contribute collectively to the significant precipitation, wind, and waves associated with a nor'easter. If the winter storms move up the Eastern Seaboard, just far enough east so that cold air stays to the west, ensuring temperatures are cold enough for snow. The Appalachia Mountains often provide a barrier that nudges the low-pressure system northeastward, keeping it over the Gulf Stream. The Atlantic Ocean and Chesapeake Bay provide lots of moisture to fuel nor'easter development and produce copious amounts of precipitation. Blocking high pressure Systems feed cold air into the storms, fueling their growth. They can also slow a storm's northward progress, increasing snowfall rates and totals, and lengthening exposure to high winds and tides. To be at their most powerful and prolific nor'easters must "hug" the coast without getting too close or too far away from the shore. Too far out and the storm's moisture field is too far away to produce much precipitation. Too close and the warm air of the Gulf Stream may change the precipitation from snow to rain (a so-called warm air intrusion). A nor'easter located about 100 miles from the shore is likely to produce the combined maximum effect of surf. beach erosion, wind speed (wave damage to roads and dwellings), wind direction (snow on back side of the storm during winter months) and cold winter temperatures. Occasionally an area of energy in the Gulf of Alaska (perhaps the remains of a Pacific storm) is the precursor for nor'easter development a week or more in the future.

An intense nor'easter off the coast of New England can produce significant, dangerous, destructive weather over a large area. Damaging winds that affect ships at sea, buildings of all sorts, air travel and road conditions can be experienced hundreds of miles from the center of the storm. For example, though centered south of Nova Scotia and east of Cape Cod, a large nor'easter may be responsible for gale force winds in Michigan, sharply colder temperatures in New York City, lake effect snow on the southern and eastern shores of the Great Lakes, and flight delays over a large area. Flight delays can affect not only domestic air schedules, but in the case of airports in, for instance, Gander, Newfoundland, can compromise the emergency landing sites in eastern Canada used for flights leaving Europe for North America. There are two meteorological phenomena that can greatly limit the life and severity of a nor'easter—the Dry Slot and Warm Air Intrusion. Both are capable of literally "shutting down" a storm. And both can occur quite quickly with little or no advance warning, making them difficult to predict. The dry slot is a phenomenon, which also occurs in hurricanes, by which dry (low dew point) air is drawn into the storm. The dry air disrupts the energy in the storm and the rotation of the cyclone, effectively shutting down the storm's snowmaking capabilities. In this case warm air is drawn off the Atlantic Ocean by the storm's counterclockwise rotation The warm air displaces entrenched colder air that fuels the nor'easter. Instead of heavy snow, a mixture of precipitation types (freezing rain, sleet, or just plain rain) is more likely to result. Snow stops and the storm is, for all intents and purposes, over.

The Inevitable Comparisons—Nor'easters & Hurricanes

Nor'easters, like hurricanes, are cyclonic lows, with winds spiraling around a central low-pressure zone. And, like hurricanes, nor'easters cause storm surge, high winds, beach erosion and high rates of precipitation. The strongest nor'easters can be equivalent to Category 1-2 hurricanes. Some years there may be multiple nor'easters, relentlessly pummeling the Jersey Shore, Cape Hatteras, Nantucket, Cape Cod and the southern shoes of Connecticut Rhode Island and Long Island with waves, wind, tides and rain and/or snow. Despite the obvious damage occurring and risks to human life associated with nor'easters, we remain meteorologically obsessed with hurricanes, even though only one hurricane makes landfall every five years or so from Georgia to New England. It is the cumulative effects from multiple nor'easters in a single year, or a period of several consecutive years, that defines the real risk and consequences of "Winter's Hurricanes". These annual combined impacts from multiple nor'easters can easily equal or exceed the damage from a single hurricane. See a detailed comparison in the table below.

Hurricane- Nor'easter Comparison

Feature	*Nor'easter*	*Hurricane*
Meteorological Type	Cyclone, with winds spiraling around a central Low Pressure zone	Cyclone, with winds spiraling around a low-pressure system zone
Where Formation Occurs	Extratropical	Tropical
Fuel for Develop	Cold air drawn into circulation	Warm air drawn into circulation
Footprint	Clouds circulating around storm center	Clouds swirling around a well-defined eye
Frequency of Landfalling	Unusual	One every five years`
Frequency of Shore Impacts	Multiple storms in a given year can cause significant cumulative damage from surge, high tides and beach erosion	None, one two per year

Of course, there are other meteorological phenomena and storm systems besides nor'easters that lead to substantial snowfalls. Next, we will look at the blizzard, prevalent in the Northern Plains, and just the mention of which brings forth visions of wind-whipped snow and white-out conditions. And we will review the quick-hitting Alberta Clipper and the downright scary and powerful Meteorological Bomb, both of which are most common over the eastern half of the United States.

Blizzards, Clippers & The Bomb

A blizzard is a major winter storm that 1) lasts for at least three hours, 2) has sustained winds or frequent gusts over 35 mph during each of those three hours, and 3) has visibilities of a quarter of a mile or less caused by falling and/or blowing snow in each of those three hours. So, the criteria for attaining blizzard status are much more detailed and prescriptive than for that of, say, a nor'easter, which is quite simply "a macro-scale, extratropical cyclone in the western North Atlantic Ocean". Blizzards are most common in the Midwest and Plains states where high winds and blowing snow are common. Falling snow is not a requirement for blizzards—blowing snow is sufficient. So, existing snow from a previous storm can even contribute to blizzard conditions in the present. And a low-pressure system per se, typically associated with most major storms, is not a necessary ingredient for a blizzard. The passage of a powerful cold front can trigger and sustain a blizzard. East Coast snowflakes and snowpacks tends to be wetter and heavier, and therefore much more difficult to render airborne. Nor'easters can occasionally reach blizzard status, but it is not common. It is here that reality and "close-is-good-enough" sometimes meet. For many, just a mention of the word "blizzard" conjures up visions of horizontal, wind-whipped snow, trapping even the heartiest of souls inside their homes, on Interstate highways closed, tractor-trailers overturned. A <u>powerful</u> storm, inspiring <u>awe and fear</u>, man completely at its mercy, lives and belongings hanging in the balance. Worrisome at a minimum, <u>downright scary</u> at its worst.

The Alberta Clipper

Although the nor'easter is Mother Nature's most prolific snowmaker in the Northeast and Mid-Atlantic states, there is another snowmaker that consistently provides snow to these same regions. And that would be the Alberta Clipper. An Alberta Clipper is a winter storm system that originates in the Canadian province of Alberta (or close by in Saskatchewan, Manitoba or even Montana). The "clipper" is named for the clipper sailing ship, famous for the speeds it was able to attain and sustain on water. Alberta Clippers get started on the western side of the Canadian Rocky Mountains. Moisture from the Pacific Ocean starts the process, and a low-pressure system (the clipper) is born.

The clipper gets caught up in the polar jet stream and travels southeastward into the northern Plains, over the Great Lakes and eventually off the mid-Atlantic coast. The atmosphere will usually have little ingrained moisture and will be travelling at a relatively high speed. Under these conditions, the clipper will bring mostly minor snow accumulations (generally 1-3 inches) to locations in its path. Higher amounts are possible but not common. Along with the quick burst of snow, a clipper generally brings colder temperatures and gusty winds. The snow it discharges is generally evenly distributed and fluffy in nature, rarely heavy. Since conditions tend to be cold when clippers move through the Great Lakes and onto the East Coast, clippers rarely precipitate anything but snow. The clippers may eventually target the Mid-Atlantic states. As they approach the Appalachian Mountains, their air masses are lifted by the geological features of the mountains. As the moisture-laden air rises, it encounters cooler temperatures. Cool air can hold less moisture than warmer air. This phenomenon leads to additional snowfall on the western side of the mountains at the expense of more populated areas of the East Coast. Very little moisture is left as the clipper crosses over the mountains and onto the Eastern Seaboard.

The Meteorological Bomb, as it is referred to by weather professionals, is a rapidly deepening low-pressure system that typically forms offshore in the western North Atlantic Ocean. A Meteorological Bomb is defined as a low pressure system with a drop in atmospheric pressure of 0.71 or more inches of mercury for more than 24 hours. The formation of the so-called bomb can be triggered by a strong pressure gradient between cold high pressure to the north and the relative warmth to the south, much like the Gulf Stream, much like the process that accelerates the development of nor'easters. And, in fact, meteorological bombs are nor'easters. Strong, upper level winds and upper air disturbances exacerbate the pressure gradient and increase the intensity of the storm.

The Clipper and The Bomb

Alberta Clippers and nor'easters are typically independent events that occur periodically from September through April in the northern Western Atlantic Ocean. But occasionally they work together. Explosive cyclogenesis is the process by which meteorological bombs are formed. Explosive development is accompanied by rapid deepening of the cyclone's low-pressure area. Alberta Clippers can be instrumental in the cyclogenesis process. To formally classify a low-pressure system as having undergone explosive cyclogenesis (or bombogenesis), a specific drop in central pressure of the low-pressure system, as a function of the latitude, is required. For example, at 60 degrees north latitude, explosive cyclogenesis occurs if the central pressure decreases by 0.71 or more inches of mercury, a remarkable deepening. Cyclogenesis is primarily a winter maritime event and is the equivalent of the rapid deepening that occurs in tropical systems like hurricanes. Bomb cyclones can produce 75 mph winds, equivalent to those of a Category 1 hurricane on the Saffir-Simpson Scale for hurricane intensity. Crédit for naming the "bomb" goes to Professors Fred Saunders of MIT and John Agyekum of McGill University in 1980.

Low pressure systems are responding to a number of critical atmospheric processes occurring on various horizontal and vertical scales in the atmosphere, including not only the pattern of upper level winds, but a whole host of other factors, including for example, diabatic heating and latent heat release from clouds/precipitation associated with the low pressure system. It's all highly non-linear; the low is responding to the sum of ALL these processes (too intricate to go into in this book), and the track and intensity/deepening of the associated low is not merely following the jet stream. The National Weather Service's Numerical Weather Prediction (NWP) model is still not getting all of the processes modeled correctly. So this is (and will be) a work in progress even as the NWS continues to make strides in NWP modeling science.

Many other processes are at play during storm development including air-sea interactions and several more well beyond the scope of this article that contribute to the explosive deepening of these lows. The extratropical transformation of hurricanes over the eastern Atlantic can be quite dramatic. Post-tropical remnants encounter a stagnant jet stream and significant rain. Many Alaskan low pressure systems arriving from the Pacific are essentially "bomb cyclones", but their effects are far less long reaching because of the sparse population of the state, meaning a much less dense population is unlikely to take note of the storm, perhaps not even aware of its arrival as compared to say a similar disturbance affecting the U.S. East Coast.

Snow can be produced under the most unlikely of conditions. Lake effect snow, a phenomenon capable of producing prodigious snowfall rates and accumulations, can occur without the need for a low-pressure system. Orographic snow is produced by rising humid air masses climbing up mountains, no matter the direction in which they are blowing. Snow squalls are intense wind-driven snow events that often precede cold fronts.

Lake Effect Snow

The phenomenon known as lake effect snow is truly one of nature's most interesting meteorological events. The ingredients required to create lake effect snow are found in only a few relatively small parts of the world. Most people have never experienced, and probably never will experience, lake effect snow. This further adds to its mystique. The process by which "lake effect snow" is produced is unlike any other. It is s a truly unique phenomenon. So, how exactly does it happen? First, we need a body of water. Evaporation causes moisture-laden air to rise above the surface of the water. Next, we need wind blowing over that body of water. The wind transports the warm, moist air, propelling it onto the land where our last ingredient, cold air, is in place. The warm moisture-laden air contacts the cold, dry air. This leads to condensation, then precipitation, and snow is produced generally, snow in the Mid-Atlantic and New England is associated with a low-pressure system, either an Alberta Clipper zipping across the Plains to the East Coast, or a major system like a nor'easter moving up the Eastern Seaboard. These storms produce snow over an area measured in many thousands of square miles. But lake effect snow is created by an entirely different process. Low-pressure system is not a necessary ingredient. Snowfall is localized in so-called "snow belts", which can be as narrow as just a few miles across. Lake effect snow is most common in early winter before the body of water responsible for evaporation has frozen over. Once that body of water has frozen there is insufficient evaporation for subsequent condensation and precipitation to occur. Wind direction and "fetch" are extremely important in determining where and how much snow will fall (more on these factors later). Other factors that can accelerate snowfall intensity include higher humidity over land, lower atmospheric pressure, wind shear and orographic effects.

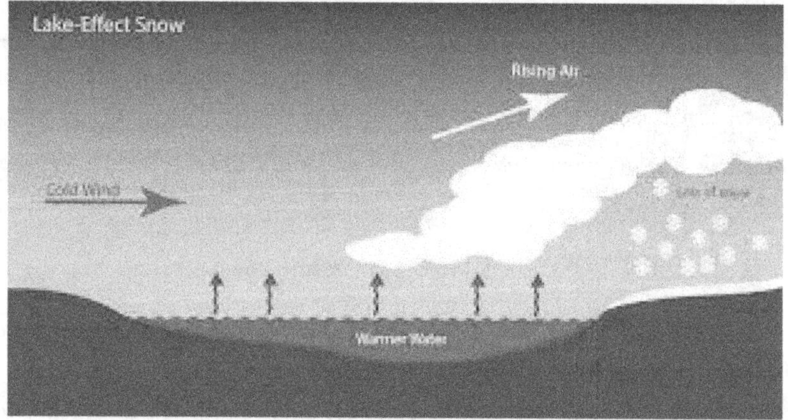

The greater the difference in temperature between the body of water from which moisture is being extracted, and the temperature of the air on the land over which it is being transported, the greater the potential for heavy snow. A major lake effect snowstorm affected Buffalo, NY and its suburbs in November of 2014 and again (even greater) in 2023. In 2014 the surface temperature of Lake Erie was 54F; at the same time a weather balloon registered an air temperature of 0.1F at an elevation of 5,000 feet above the lakeshore. This significant difference was a primary reason for the severity of the November 2014 Buffalo lake effect snow event. Lake effect snow is most likely and most severe in late fall and early winter when the lake water is relatively warm but cold air temperatures are increasingly likely. The first of the Great Lakes to freeze is generally Lake Erie, the shallowest of the lakes. As winter wears on, both the air and water temperatures drop. The driving force for heavy snow is reduced. Eventually, the body of water may freeze over, and the "lake effect snow machine" is effectively shut down.

Here is some of the meteorology associated with lake effect snow:

A temperature difference of 23F between the lake temperature and the height in the atmosphere at which the barometric pressure measures 850 mB provides for absolute instability and allows vigorous heat and moisture transportation verticality.

Fetch is the distance travelled by wind across open water. Fetch is a surprisingly important factor in determining where and how much snow will fall during a lake effect snow event. A fetch of many miles (60 or more) is required to create significant lake effect snowfall. This distance is necessary for the air to become sufficiently saturated with moisture on its journey across the open water. Some places, like Buffalo and other locations on the eastern shore of Lake Erie, are optimally positioned to receive a fully saturated air stream ready to discharge its moisture in the form of snow once land and cold air are encountered. All other factors remain constant; the biggest "lake effect" snow events occur when the wind is blowing parallel to the length of the lake. Even a slight change in wind direction will move the location of maximum snowfall to a new spot. In the record-breaking Buffalo lake effect snow event of 2014, the amazing (and unusual) persistence of wind direction and cold air at the shoreline led to the exceptionally heavy snow experienced. Lake effect snow systems also create their own micro-climates. As snow begins to form, air rises in snowy regions and sinks on its sides, creating a miniature area of high pressure. The result is generally clear and distinct boundaries between areas receiving heavy snowfall and those, potentially just across the street, receiving none.

Many bodies of water around the world experience some or all of the factors that lead to lake effect snow.

But no bodies of water "do it as well" as the Great Lakes. So, yes, it's the Great Lakes' propensity for snow, and lots of it, from this unique phenomenon, that has led to the term "lake effect snow". Portions of all the states bordering the Great Lakes receive lake effect snow most years, though in varying degrees of frequency and intensity. These states are Minnesota, Michigan, Wisconsin, Illinois, Indiana, Ohio, Pennsylvania and New York. Because the prevailing winds that bring cold early winter air to the Great Lakes region are predominantly from the northwest, it is the eastern and southeastern (one might say opposite) shores that receive the bulk of Great Lakes' lake effect snow. So, South Bend, Indiana gets significantly more lake effect snow than Chicago; Erie, Pennsylvania more than Milwaukee, Wisconsin, and so forth. The shores of Utah's Great Salt Lake, the Baltic and Black Seas, and the Sea of Japan, among others, can also receive lake effect snow. The East Coast of the United States sometimes receives snow from a lake effect-like phenomenon. Easterly winds blowing over the relatively warm waters of the Atlantic Ocean and Chesapeake Bay may encounter cold air as they reach the shore and create "lake effect like snow". But this tends to be more of a meteorological curiosity than a phenomenon of concern. Snow added to the primary weather-maker is small if at all.

Lake effect snow is unlike any other snow-producing meteorological phenomenon. Not associated with a storm system, it is like a weather orphan, doing its own thing while the rest of the country watches, enthralled and fascinated, but not concerned. After all, this snow system does not need to be watched like an advancing nor'easter. No, it is going to stay basically where it is, on the shore of the lakes, perhaps moving several miles along the shore but posing no real threat those outside the immediate snow belt areas.

The Derecho

The National Weather Service defines a derecho as "a fast-moving, long-lived, large, violent thunderstorm complex that creates wind damage along a swath at least 240 miles wide while travelling at a forward speed of at least 58 mph". Yikes. Most people have never even heard the term. This is obviously a strong storm capable of creating substantial damage. Perhaps should come as no surprise, given the extent of these requirements, that there are so few of these storms. <u>Derechos are distinguished from thunderstorms and tornadoes because their winds are predominantly "straight-line".</u> By comparison, thunderstorms and tornadoes are composed primarily of rotating winds. The damage inflicted by a derecho is equivalent to that of a weak-to-moderate tornado. And, because the damage occurs over an area hundreds of miles wide, the cumulative damage created by a derecho can actually be much greater than that caused by a tornado. The very high straight line winds associated with the leading edge of a derecho are associated with the forward movement of the air mass itself, creating wind speeds at least equal to the derecho's forward speed. Additional energy may be associated with the storm itself. Statistically, one derecho is expected to affect the Washington D.C. Metro area about every four to five years. Actually, there has been only one since the turn of the century.

The Derecho of 2012 originated west of Chicago late in the morning of Friday June 29, 2012. It raced eastbound through Fort Wayne, Indiana just after noon. Seven hours later, still a formidable force and still intact, it arrived in Northern Virginia. A gust of 79 mph was registered in Reston, VA. The derecho inflicted significant damage over a huge area, mostly in the form of fallen trees that toppled telephone poles, took down power lines and fell on people's homes, businesses and cars. The storm stayed together for a remarkable 900 miles. It claimed 22 lives, damaged over 1,000 structures and led to power outages affecting nine million customers.

The Atmospheric River

Increasingly of concern along huge tracts of land stretching from San Diego northward through Los Angeles to Seattle and Portland, including the mountainous terrain of the Sierra Nevada and Northern Rockies clear through the states of Oregon and Washington is the new meteorology term "Atmospheric River." Much like the moniker Polar Vortex was when it first came on the scene several decades earlier, the atmospheric river is an apt terminology to describe the path of a visible river of water vapor that rushes toward the U.S. West Coast, whose mountains eagerly await the arrival of the water vapor and the consequent condensation of untold trillions of gallons of rainfall and, at altitude, snow. Low pressure systems get caught up in the "river" and, with the path virtually unchanged, the same areas get socked by the storms more than once.

February and March 2023 featured one of the strongest and longest-lasting editions of this phenomenon. Many higher altitude locations endured seemingly endless snow events. South Lake Tahoe had once in a lifetime snow totals of 10 feet of more. On the warm side of the events, rain kept on going and going. Eventually the National Guard had to be brought into the region to help the resident get out of doors. On the warm side of the storms, rainfall amounts were in some obscure locations measured in feet. Los Angeles and San Diego even got into the action. Despite the damage that occurred, there is a silver lining to sch an event; the rain and snow was so heavy, and the coverage so great that the drought that had been around for what seemed like forever is suddenly gone!

CHAPTER 5 — CLIMATE

Weather is what one is experiencing right now. Temperature, precipitation, wind, humidity. It is the meteorological state of the atmosphere at a given place and time on the Earth's surface. In many places and parts of the world, weather is highly changeable week-to-week, day-to-day, hour-to-hour.

Climate is the accumulation of data resulting from weather that has already happened. It is the composite of day-to-day weather over long periods of time. Climate represents an area's long term weather. So, the light snow and cold windy weather you experienced today will contribute to the databases from which climates are defined. The National Weather Service and others consider the 30-year average to represent a reasonable time with which current and past meteorological data can be compared. While weather is highly variable and often unpredictable, significant changes in climate only occur after many years, usually thousand and millions. Research is showing that "significant" changes can occur in local climate in periods of less than one hundred years.

The difference between climate and weather can be summarized by the very instructive phrase "Climate is what you expect. Weather is what you get."

Every point on the Earth's surface has a climate. There are many factors that determine what that climate will be. These factors are numerous, and their interactions are complex and not necessarily well understood. But generally the following factors are most instrumental in creating the Earth's unique set of distinct climate zones

- Latitude on the globe – in general the farther from the equator, the colder the surface temperature.
- Elevation – climate zones are generally aligned with altitude, with higher altitudes being colder and receiving more snow.

- Proximity to the ocean or a large lake – since water temperatures are more stable and slower to change, large bodies of water can keep adjoining land from becoming either exceptionally hot or exceptionally cold.
- Topography – mountainous terrain can facilitate orographic precipitation by helping to lift moist air to cooler regions of the atmosphere where condensation occurs.
- Vegetation – absorbs, retains and transpires water and is a surprisingly important participant in the water cycle.

Over the years there have been many attempts to use a set of parameters to define the world's climate. In the tenth century, Chinese scientist Shen Kuo was one of the first to take note of what we now refer to as climate cycles. After observing petrified bamboo at a location not then suitable for growing the plant, he correctly theorized that climates shift naturally after very long periods of time. The ancient Greeks believed that weather and climate were directly related to a given location. While latitude is a very important factor in determining climate, it's not the only one.

Major advances in understanding and characterizing the Earth's climate occurred in the late 1800s and early 1900s. Wladimir Koppen, a Russian-German climatologist, is credited with developing the first comprehensive climate classification system, which bears his last name. The challenge for climatologists has always been to create a limited classification system that 1) has a sufficient number of separate climate zones that are clearly distinct from each other, and 2) which in total characterize the broad diversity of the Earth's weather. There is no single set of climate zones that is used exclusively by climatologists, nor in fact does there need to be. A common shortcoming of climatology zone designations is that they contain distinct boundaries between the zones they define. A gradual transition of climate properties between adjacent zones is more typical of what occurs in nature.

The Koppen Climate Classification System

The Koppen Climate Classification system is based primarily on average annual and monthly temperature and precipitation. It is the most widely used system for classifying the world's climate. This is due mostly to the relative simplicity of the system, including the use of words and concepts that are clear and relatively easy for many to understand

Although early versions of the Koppen system contained some omissions and deficiencies, Koppen's climate classification zones were remarkably accurate considering his limited and slow transportation options (boat and horse among them) and rudimentary meteorological tools and instruments that were at his disposal in the late 1800s to visit, record and evaluate climate systems across the globe.

The Koppen Climate Classification system has five primary climates:
- Tropical
- Dry
- Temperate
- Continental
- Polar

These primary climate classifications can be further divided into secondary classifications.

Many of these terms are ones even laymen not trained in meteorology can probably describe quite well. Classifications in each column are roughly aligned with each other.

Rain Forest	Humid Continental	Subarctic
Monsoon	Oceanic	Tundra
Tropical Savanna	Mediterranean	Polar Ice Cap
Humid Subtropical	Steppe	Desert

In total, the Koppen system has twenty-nine distinct climate zones. Seventeen of these zones characterize the climate of some part of the 50 United States. Some zones cover a significant portion of the land mass of the U.S.;

others a much smaller portion. The following table provides some examples of the Koppen climate classifications that exist in the United States.

Classification	Climate Description	Where
Humid Subtropical	Warm, moist air flowing from the tropics creates hot, humid summers with significant rainfall from frequent thunderstorms.	Southeastern U.S. (Nashville, New Orleans, Orlando)
Hot Summer Continental	The average temperature is above the warmest months and below 25F in coldest.	Northeastern U.S. (Chicago, Boston, Minneapolis)
Mediterranean	Hot, dry summers with some fog along the coast; winters changeable with rain and moderate temperatures.	U.S. West Coast (San Francisco, Los Angeles)
Continental Subarctic	Little precipitation. Monthly temperatures exceed 50°F for one to three months of the year. Permafrost is prevalent. Winters include up to six months of temperatures averaging below 32 °F.	Much of Alaska (Fairbanks)
Tundra	The warmest month averages less than 50F. Occurs at northernmost edge of continents or at high altitude above the tree line.	Barrow, Alaska

Tropical Monsoon	Has an exceptionally rainy season that lasts for several months and is ushered in by moist prevailing winds.	Miami

There are several climates that do not exist anywhere in the 50 United States (for this I think we can be grateful!).

Tropical Rain Forest	Characterized by annual rainfall greater than 69" and mean monthly temperatures exceeding 64 °F during all months of the year.	Samoa, Congo, Philippines, Singapore
Bicontinental Subarctic with Severe Winters	Temperatures in coldest month lower than minus 40F.	Only in eastern Siberia

North America map of Köppen climate classification

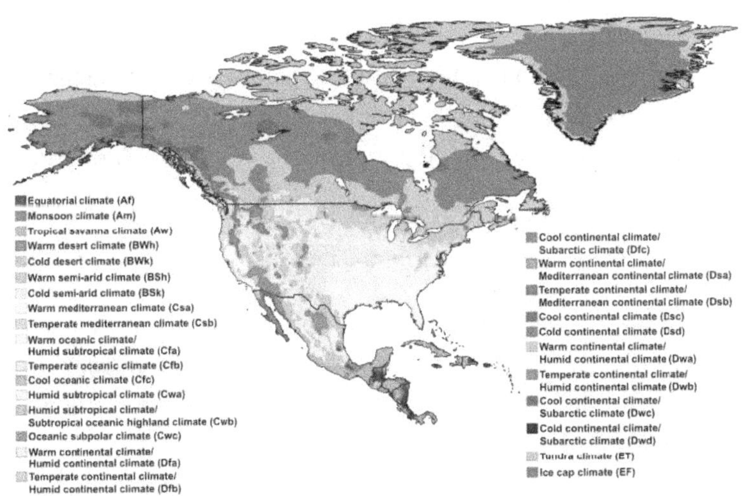

Among other climate classification systems is the Bergeron Air Mass Classification system named for Tor Bergeron, a Swedish meteorologist. It is the most widely accepted climate classification system whose zones are based ai air mass movements. It portrays the moisture properties of the air mass (e.g., Continental Dry or Marine Wet) and the thermal characteristics of the air (e.g., Tropical or Polar). American geographer Glenn Thomas Trewartha modified Koppen's system in the 1960s to address a number of mid-latitude deficiencies. Trewartha's system incorporates living organisms on the basis that native vegetation is the best expression of, or result from, climate. Vegetation type, location and density are combined with other climatological characteristics to create the Trewartha Climate Classification System

Microclimates

A microclimate is a local set of atmospheric conditions that differ markedly from those of nearby surrounding areas. Microclimates exist, for example, near bodies of water which may cool the atmosphere during the day from a phenomenon known as "onshore breezes".

Microclimates also occur in heavily populated urban areas where brick, concrete and asphalt absorb and retain heat from the sun's energy. This creates a so-called "Urban Heat Island" effect that can be as much as 6-8 degrees Fahrenheit warmer than relatively nearby rural areas. The San Francisco Bay area is a region well known for its microclimates, which have a wide range of temperatures within just a few miles of each other. This is due primarily to the influence of topography on the circulation of marine air.

Climate Data

There is a lot more to climate than slicing the earth's surface into climate zones. Much of what we do with climate involves decisions made on a much more local level, such as where we live, our travel destinations, and how we live.

For this we need more specific information; we need data.

Shown directly below is the meteorological data collected at Dulles Airport in Northern Virginia. The data include not only what you expect, like high and low temperatures, rainfall and wind speed, but many more categories like how cloudy or sunny the day was, how many degree days were accumulated and the relative humidity.

```
WEATHER ITEM    OBSERVED TIME   RECORD YEAR NORMAL DEPARTURE LA
ST
                VALUE   (LST)  VALUE        VALUE  FROM       YEAR
                                                   NORMAL

.............................................................
TEMPERATURE (F)
 YESTERDAY
  MAXIMUM        69  12:50 PM  92    1990  71     -2         68
                                     2009

  MINIMUM        38   2:38 AM  31    1967  47     -9         53
  AVERAGE        54                        59     -5         61

PRECIPITATION (IN)
  YESTERDAY      0.02          0.93  2004  0.12   -0.10      0.03
  MONTH TO DATE  0.83                      2.95   -2.12      2.25
  SINCE MAR 1    2.40                      6.45   -4.05      4.17
  SINCE JAN 1    6.03                      12.00  -5.97      10.37

SNOWFALL (IN)

  YESTERDAY      0.0           0.0   2001  0.0    0.0        0.0
                                     2002

  MONTH TO DATE  0.0                       0.1    -0.1       T
  SINCE MAR 1    T                         4.0    -4.0       3.5
  SINCE JUL 1    0.4                       21.0   -20.6      15.8
  SNOW DEPTH     0


 DEGREE DAYS
  HEATING
   YESTERDAY        11                     7      4          4
   MONTH TO DATE   185                     293    -108       288
   SINCE MAR 1     755                     940    -185       811
   SINCE JUL 1    3709                     4450   -741       3865

  COOLING
   YESTERDAY        0                      1      -1         0
   MONTH TO DATE   44                      15     29         11
   SINCE MAR 1     45                      17     28         14
   SINCE JAN 1     45                      17     28         14
.............................................................
+

WIND (MPH)
  HIGHEST WIND SPEED    18   HIGHEST WIND DIRECTION    S (200)
  HIGHEST GUST SPEED    24   HIGHEST GUST DIRECTION    S (200)
  AVERAGE WIND SPEED    5.6


 SKY COVER
  AVERAGE SKY COVER 0.7


WEATHER CONDITIONS
THE FOLLOWING WEATHER WAS RECORDED YESTERDAY.
  NO SIGNIFICANT WEATHER WAS OBSERVED.


RELATIVE HUMIDITY (PERCENT)
  HIGHEST    86          5:00 AM
  LOWEST     41         12:00 PM
  AVERAGE    64
```

Finally, we have the "one off" type of data. Rather than trying to explain what this data is, and how and when it is gathered, take a look at the examples below. Data are grouped by frequency of gathering, which ranges all the way from continuously to once a year or less.

<u>Periodically</u>
- Atmospheric temperature, humidity and barometric pressure gathered from weather balloons launched from Local Weather Forecasting Offices LWFO) daily at 7 a.m. and 7 p.m.
- Snowpack at a given location in the Sierra Nevada Mountains of California
- Each year's dates for peak blooming of the cherry blossoms at the Tidal Basin in Washington D.C. (responsibility of National Park Service)
- Moisture levels in the rich topsoil of northern Indiana

<u>Continuously</u>
- Pacific Ocean water temperature at a buoy floating 445 miles southeast of Honolulu
- The number of days with an inch or more of snow on the ground at the Blue Hill Observatory outside Boston, MA
- The depth of Lake Mead behind the Hoover Dam
- Maximum and minimum temperature, rainfall and snow accumulation and humidity and barometric pressure at all Local Weather Forecasting Offices, and from Automated Surface Observing Stations. The 10,000 or so Cooperative Weather Observers collect the same data except for barometric pressure and relative humidity.

<u>During Specific Events</u>
- Hurricane Hunter aircraft data on storm direction, barometric pressure and wind speed gathered during flights to and inside tropical storms and hurricanes every six hours or so.

Once

- Inches of beach sand eroded at specific Jersey Shore locations following the passage of a nor'easter
- Tree ring pattern on a 150-year-old fallen redwood tree
- Actual landfall location of a hurricane compared to the predicted landfall locations 24, 36, 48 and 72 hours earlier.
- Path, extent of damage and estimated strength of a tornado long since expired.
- Total storm snowfall depths after the snow has stopped.
- Record daily low and high temperatures.
- Extent of forest fire damage after the fire has been extinguished.
- And that list constitutes just a portion of collected climate data.

Timeframes

Climate is commonly defined as the average weather over a long period of time. For many climatological purposes, this can be assumed to be hundreds or even thousands of years. But when comparing day-to-day temperature and precipitation data with historical climatological data, the standard averaging period is only 30 years. Thirty years is long enough to smooth over extreme, record-breaking daily high or low temperatures, and short enough to show longer climatic trends.

A commission, authorized by the International (now World) Meteorological Organization (GMO), ruled its 1934 meeting in Wiesbaden, Germany that the thirty-year period from 1901 to 1930 would be the reference time frame for climatological standard normals. Subsequently, every 10 years thereafter, a new 30-year average will be calculated and used for climate comparisons. The concept of a 30-yr average has withstood the test of time and is still used extensively today.

Climate Data

Current day climate data is captured and recorded using modern weather instruments like thermometers, barometers and anemometers that generally provide direct, precise, quantifiable and reproducible measurements. Included in the array of "facilities" providing data are weather balloons, buoys, satellites, local weather forecasting offices (LWFO), unmanned automated surface observing stations (ASOS), unmanned land-based coastal observing stations, weather surveillance radar and cooperative weather observers (people).Accurate measurements such as these have been possible only since the late 1800s, a period of about 150 years, insignificant when considered in the context of the Earth's age of 4.5 billion years.

The volume of climatological data collected every day is staggering—who knows how many millions of individual data points are added to databases every year. Climatologists study, assess, integrate and make this data understandable so society can plan its activities, design its buildings and infrastructure, and anticipate the possible effects of a changing climate on their lives.

There are also relatively obscure, yet meticulously maintained, daily record high minimums, and their companions the daily record low maximums. Although of some interest to weather geeks, these parameters don't generate the level of interest associated with the more traditional and more easily understood daily record low and daily record high temperatures. However, when analyzed in its entirety, a database of record high minimums (as obscure as they might appear) can provide a glimpse at patterns of very cold, very warm, very wet or very snowy weather.

Climatological data allows radio and TV weather reporters, newspapers and social media to provide their listeners and viewers with some perspectives regarding either an immediate past day's weather, today's weather, or the upcoming forecast.

- For instance, it's not uncommon for the "5-Day Look Ahead" TV graphic to show not only the high and low temperatures forecast for that period, but also the "average" for each of the five days. This is not a formal NWS outlook.
- Or you might hear "It's been an exceptionally wet month so far and we are expecting more rain over the next few days. We've already had nearly five inches of rain this month and we average just a little more than three."

But there are far more uses for climatological data than just keeping track of records and averages, and comparing them to more recent events.

What Do People like Us Do with Climate Data?

Every day, knowingly or not, people use climate information to help them make important decisions like where to live, where to go on a skiing vacation, when to go to the Caribbean, when to plant flowers and vegetables. If you are thinking about moving to a part of the country you have never lived in, or maybe never even visited, climate data can be a big help. Each person has his or her own views about specific climates. Some kinds of weather (low humidity, copious snow) may be viewed as highly desirable by some of us, but not nearly as sought-after by others.

Here are some weather phenomena and some cities that are among the top ten in their annual occurrence.

Weather Parameter	Cities
Warmest Annual Average Temperature	Key West, Miami, West Palm Beach
Coldest Annual Average Temperature	Duluth, Fargo, Bismarck
Largest Annual Temperature Variation	International Falls, Williston, Aberdeen
Highest Annual Rainfall	Mobile, Tallahassee, New Orleans
Most Days in a Year with Rain	Syracuse, Buffalo, Erie
Fewest Days in a Year with Rain	Santa Barbara, Long Beach, Great Falls
Driest (lowest annual rainfall)	Yuma, Las Vegas, Bakersfield
Snowiest	Marquette, Syracuse, Flagstaff
Sunniest	Phoenix, Tucson, Reno
Cloudiest	Seattle, Portland, Binghamton
Most Humid	Port Arthur, Lake Charles, Corpus Christi
Windiest	Amarillo, Cheyenne, Boston
Most Thunderstorms	Fort Myers, Tampa, Orlando
Most Landfalling Hurricanes Making	Cape Hatteras, Delray Beach, Boca Raton
Most Hail (frequency and severity)	Wichita, Denver, Shreveport
Most Tornados	Oklahoma, Tulsa, Dallas
Least Weather Variety	San Francisco, San Diego, Los Angeles

Frost Free Days

Every spring, as we slowly make our way out of winter and into the first truly warm days in months, there is a strong desire to begin that most pleasurable rite—planting our favorite flowers and herbs. And indeed weather in March in, for instance Northern Virginia, can be warm enough for long enough to convince you to head to the nursery, get out the shovel and start planting. But before we do, let's take a look at some climatological data from Fairfax, VA, USA for March that does an excellent job of illustrating the impacts of climate variability.

The five days March 7-11, 2016 had an average daily mean of 60F, 18F above normal, and daily high temperatures that included 80F on March 9 and 78F on March 10. Then, just 10 days later, lows of 31.5F on March 21 and 28.7F the next day. One needs to use climatological data and the so-called "Freeze/Frost Occurrence Data" to guide us in determining appropriate times to plant frost-vulnerable annuals. The table below (for Dulles) provides guidance on when to plant so-called "tender vegetation".

Date	Norm Daily Mean	Norm Hi	Norm Lo	Record Hi	Record Lo	Probability for 28F
March 26	48	59	36	78	17	90
April 7	52	64	40	91	21	50
April 20	56	68	43	88	28	10

The table tells us there is a 90% probability there will a temperature of 28F or lower after March 26, a 50% probability such a freeze will occur after April 7, but only a 10% probability of cold weather of this magnitude after April 20.

The United States Department of Agriculture (USDA) maintains the "USDA Plant Hardiness Zone Map" and associated information. The map is the standard by which gardeners and growers determine which plants are most likely to thrive at a given location. The interactive map is of course based on climatological data, which in this case is the average annual minimum winter temperature. The map is divided into 16 zones, with each zone representing an area having the same 10 degree Fahrenheit zone.

Infrastructure – Climate Considerations

Rainfall directly affects a road's storm water collection, removal and retention systems. If both Phoenix and Fairfax, VA were building an interstate highway, Phoenix would actually have larger drains and a larger retention pond because the data show Phoenix typically has at least one or more cloudbursts each summer that can produce rainfall rates of 2-4" per hour, perhaps for only 20-30 minutes. From a design perspective, the drains and retention ponds will be sized based on the short term rainfall data rate in addition to the total annual totals. One might intuitively expect there would be larger and more costly rainfall collection and removal systems in Fairfax County, VA, USA (annual rainfall of 40-45") than in Phoenix (annual rainfall 5-8" per year).

In snow-covered Vermont, Interstate highways and other major roads are designed with extra-wide shoulders to allow monster snowplows to push the snow completely off the highway's travel lanes. Swales and other imaginative land contouring designs allow the machines to also blow, throw or push the snow completely off the road (including the shoulder) without encountering any earthen or man-made walls. This added shoulder width and "free-throw" space is expensive, but necessary, in snowy climates. However, such designs perhaps cannot be justified in Northern Virginia based on climate data which shows that a major snowfall of more than say a foot occurs on average every four to five years.

Building Houses Based on Climate

Climate is a major determiner for a myriad of home building requirements, everything from air conditioning to insulation to hurricane protection. Our houses are (or should be) designed based on the prevailing climate where they are built. Climate data is used extensively to provide the basis for many design parameters. The four elements of climate that have the greatest influence on building design are temperature, humidity, sun, and wind. Temperature, of course, can be both a liability and a benefit. A liability in warm climates if housing designs allow relentless sun exposure that quickly heats the structure, making cooling difficult and costly. A benefit is provided in colder climates where that same southern exposure can provide welcome, sought-after heat. Throughout the years, mankind's response to climate conditions led builders to create architectural structures that took advantage of positive climatic features in a home's design, and minimized liabilities that were unwelcome. This led to the development of regional architectural styles.

Where	What	How
New England	Deflect harsh winter winds and shed roof snow loads	Saltbox homes with long sloping roofs
New Mexico	Ease wide and uncomfortable with daily temperature swings	Adobe construction that heats and cools slowly
Charleston, SC	Prevalent hot, humid conditions	Breezy verandas
Nebraska	Insulation against the relentless bitter arctic winter winds of the Great Plains	Early settlement sod homes

Today home builders and designers have powerful and versatile tools and models to help them address a myriad of home climate-oriented architectural issues such as type of roofing materials and associated drainage systems (asphalt, metal, gutters and downspouts), options for heating including heat pumps, gas fired furnaces and/or electric radiant heat floor tiles, type and thickness of insulation for attic applications. Window design (e.g., double or triple glazing), use of cross-ventilation, attic fans, ceiling fans, and/or other air movement equipment. Coming up next is a shortened version of a major undertaking to define and describe the diversified climate of Virginia's beautiful Commonwealth.

The Climate of Northern Virginia

The climate of Northern Virginia is temperate—that is, it lies geographically between the tropics and the polar regions such that extreme cold and heat are rarely experienced. There are four distinct seasons, each with its own special weather. Summer is hot and humid, fall is cool and refreshing, winter has cold and snow but is not harsh. Annual rainfall is plentiful, and generally occurs with regularity on about 110 days throughout the year. Snow can be expected every winter, although annual snowfall amounts vary widely. The highest ongoing weather risk is from thunderstorms that are numerous and oftentimes severe. Major snowstorms, tornadoes, drought, ice storms, Derechos and inland effects from landfalling hurricanes do occur, but only infrequently and not to the level of severity seen in other parts of the country for these same events. More than half the days in the year (about 200) are considered sunny.

Northern Virginia is located in a meteorologically dynamic region of the Earth. Changeable day-to-day weather is the norm, and one can expect to see remarkable differences in temperature, precipitation, humidity and wind speed and direction within almost any two-week period. Northern Virginia is located in a geographically diverse region of the Earth's surface, and its climate is uniquely determined by the presence of, and interactions between, the Shenandoah Mountains to the west, the Atlantic Ocean and Chesapeake Bay to the east, the Gulf Stream lurking offshore in the Atlantic and a broad "plain" that runs southwest/northeast from roughly Wilmington, DE to Richmond, VA with only minor changes in elevation. Prevailing summer southwesterlies have little geographic resistance as hot, humid air is pumped up from the Southeast; similarly, this "channel" provides a generally smooth path for the cold and snow associated with nor'easters to barrel into Northern Virginia.

Geographic Feature	What It Does	Primary Impact on N Virginia
Shenandoah Mountains	Facilitates Orographic Precipitation	Limits snowfall from Alberta Clippers
Shenandoah Mountains	Facilitates Cold Air Damming	Freezing Rain and Ice Storms
Shenandoah Mountains	Facilitates Adiabatic Warming	Higher air temperatures on down sloping sides of the mountains
Gulf Stream	Conveys storms up the Atlantic Seaboard	Increased snow, wind, rain from Nor'easters
Atlantic Ocean	Provide warm, moisture laden atmosphere	Enhances convective energy & strength of storms
Atlantic Ocean	Generate persistent moist, easterly winds	Prolonged periods of cloudy, damp weather
The Space Between	Facilitate unimpeded movement of air masses	Hot humid summers; greater impact from winter storms

Rain is plentiful and some accumulation can be expected every month.

- Total annual rainfall averages about 42".
- Average monthly rainfall ranges from 2.61" in January to 4.72" in May.
    o There is a wide variation in annual rainfall totals year-to-year.
- Totals range from 27.02" in 2007 to 65.67" in 2003.
    o There is no distinct wet or dry season as occurs, for stance, in California.
    o Measurable rainfall (0.01" or more) falls on about 110 days a year.
    o Very heavy rainfall occurs periodically from thunderstorms, tropical systems and nor'easters.

When the Earth Got Cold

Throughout history the Earth has gone through several cooling periods, some of short duration, some lasting many centuries. Some caused minor inconveniences for humankind, like the "Year with no Summer". Some resulted in glaciers that covered large portions of the Earth's surface. In general, the cause(s) of these temporary cooling periods are not fully understood, nor are the reasons why each time it cooled the Earth stopped cooling and got warmer again. Let's look at a few of the times when the Earth got cold.

For a Few Years

As March of 1816 turned into April over the Northern Hemisphere, as longer days took hold, there was the eternal anticipation of a warm summer and, later on, a bountiful harvest. However, as May unfolded, weather events more typical of late fall and early winter were, instead, being experienced, and as the summer of 1816 wore on, weather unlike anything ever experienced was becoming a source of real concern. Much lore has accompanied the weather that happened 200 years ago, the so-called "Year without a Summer", the summer of 1816.

As it turns out, much of the lore of that summer's weather was just that, lore. It didn't snow in South Carolina, Long Island Sound was not iced over, New England was not covered with snow. On the other hand, there were some remarkable meteorological events and features (now verified) that occurred that summer. The summer of 1816 was the coldest or second coldest on record, routinely five or more degrees below normal in many places, including Philadelphia and much of Connecticut and Massachusetts. Although not extraordinarily below average when looked at over a 90-day summer, there were remarkable, indeed unprecedented, summertime intrusions of very cold air into the Northeast and Mid-Atlantic states (and Northern Europe) that affected agriculture, and confused and worried residents. Killing frosts occurred in mid-May from New England to Virginia, starting in May a series of cold air masses have invaded New England. Up to a quarter inch of ice covered vegetation on May 29 in Erie, Pennsylvania. A nor'easter whitened the ground of southern New England on June 6 and led to snowfalls of 6-12 inches over much of northern New England several days later

For a Few Decades

Quite unexpectedly much of the United States experienced a period of significantly colder than normal temperatures in the 1960s and 1970s. At Reagan Airport just outside Washington, D.C., with records dating back to 1871, six of the ten coldest winters in the past 70 years occurred between 1961 and 1978. There was real concern about global cooling.

Interestingly enough, some of the same worries that we hear today about global warming were being discussed then in response to the noticeably colder temperatures being experienced at that time. "There are ominous signs that the Earth's weather patterns have begun to change dramatically, and that these changes may portend a drastic decline in food production," To scientists, these seemingly disparate incidents (*tornados, typhoons, droughts*) represent the advance signs of fundamental changes in the world's weather. In the early 1980s, more normal temperatures returned and the global cooling phenomenon, and it's associated predicted dire consequences, were just another blip in meteorological history.

For a Few Centuries

Between roughly 1300 and 1870, much of Europe and North America experienced much colder temperatures than had occurred for many centuries before. Unlike true ice ages, these colder than normal years were composed of distinct periods of especially cold weather that affected specific regions of the northern hemisphere, rather than simultaneous global-wide periods of extreme cold marked by increased glacier formation. The period from 1300-1870 is referred to as the Little Ice Age. The Little Ice Age followed the Medieval Warm Period which occurred between roughly 950 and 1250.

The literature provides an array of dates representing the "coldest" periods associated with the Little Ice Age, many of them overlapping and none of them consistent. One resource claims that there were three particularly cold periods beginning around 1650, 1770 and 1850; another says that especially cold weather occurred from 1600 to 1780; a third "pinpoints" the coldest periods as occurring from 1420 to 1510 and from 1670 to 1720. No matter what the exact dates, cold periods were invariably separated by periods of warmer than normal temperatures. And, regardless, the colder than normal temperatures had profound short and long term implications for Earth and its people. Growing seasons were shortened remarkably, perhaps by 1-2 months in England at one point. Over the course of multiple centuries, the make-up of entire European forests simply changed. Initially composed primarily of beech trees, they slowly transitioned to first oak and then pine. Wine production in southern England, which had flourished between 1100 and 1300, was gone by the 1400s.

Of course, there weren't just "losers" in these periods of colder weather, a concept that needs to stay vital in the current period of global warming. Fishermen benefitted from colder ocean temperatures that helped restock the over-fished Newfoundland Banks. English fishermen found great joy in the southward migration of herring from the coasts of Norway. And the failure of crops in Norway, coupled with diminished fish stocks, led to the growth in ship building and merchant shipping in that country, initially for importing food but later for all sorts of cargoes. Timber to support ship building was grown and prospered on land too cold to support crops.

More frequent and severe storms were reported during the Little Ice Age, attributed primarily to the repositioning of the polar jet stream. Cold winters during the Little Ice Age also created some interesting meteorological/societal tidbits, each of which is only a dream when viewed in the context of today's weather in those locations. Canals and rivers in Great Britain and the Netherlands froze repeatedly, allowing annual ice skating and winter festivals to take place. In 1622 the Bosphorus (current day Istanbul) froze over. In 1658, the Swedish army crossed the Baltic Sea to attack Copenhagen (a bridge now makes that trip possible at any time). In 1608 explorer Samuel Champlain reported ice on Lake Superior to be thick enough in June to support a man's weight. In 1780 ice in New York Harbor was thick enough for people to walk between Staten Island and Manhattan.

Historical accounts suggest that particularly massive volcanic eruptions occurred in 1258, 1268, 1274 and 1284 at various locations around the globe. The repeated releases of ash and sulfur dioxide occurring so close together may have negated the usual, several-year "return to normal" that occurs after individual eruptions and could have set in motion a cooling process that initiated the Little Ice Age. Disruption of the Earth's thermohaline circulation caused when large amounts of fresh water introduced into the Atlantic Ocean from the melting of Greenland's ice cap during a period of global warming may have slowed circulation of the Gulf Stream and contributed to global cooling.

A decrease in human population, in some cases dramatic, for instance the Black Plague in the 1340s, may have led to significant reforestation. Also there's a possible connection with the "Great Dying" the genocide of indigenous peoples across the Americas from colonialism and associated spreading of diseases in the 16th century and cooling (causing an increase in wild plants as formerly cultivated fields turned to native plants that produced more oxygen and thus caused a dip in CO_2 levels). Additional vegetation would in turn consume significant amounts of carbon-based greenhouse gases and contribute to global cooling (just the opposite effect caused by today's rampant deforestation).

For Epochs – The Ice Ages

We've looked at climate changes that lasted a few years, a few decades and centuries. How about changes that lasted a really long time—really long, like millions of years. So long in fact that we have to introduce new terminology just to put these long timeframes in perspective.

An epoch is a subdivision of the geologic timescale that is longer than an age but shorter than a period. An *ice age* is a period of long-term cooling that results in the presence or expansion of polar ice sheets and alpine glaciers. Within any given ice age, individual periods of especially cold weather are known as *glacial period*. Intermittent warm periods are known as *interglacial periods*. The most recent glacial period occurred during the Pleistocene Epoch of the Quaternary Glaciation. Commonly called just "The Ice Age", it lasted about two million years, ending around 10,000 years ago. Counted among its "achievements" are the Greenland, Arctic and Antarctic ice sheets (still present today, though perhaps fleetingly so). We are currently living in an interglacial period of the Holocene Epoch. Geological, chemical and paleontological evidence are used to define periods of the earth's history when ice ages were present. *Geological* evidence results from rock scouring, glacial moraines and valley cutting. **Chemical** evidence is obtained by comparing the presence, absence and ratio of various isotopes that provides data to construct a temperature record.

Ice cores provide invaluable evidence regarding the presence and timing of ice ages. There have been at least five ice ages in the history of the earth. The earliest, called the Huronia, occurred about 2 billion years ago. The most severe glaciation in the last billion years occurred from 850 to 630 million years ago. The cooling and glacial advance were so great that ice sheets may have reached the equator.

Not fully understood are the causes that trigger temperatures cold enough to cause glaciers to grow and move beyond their normal locations. It is known that consistent temperatures about 12F below average are sufficiently low to prohibit significant summer snow melt in the arctic, thereby ensuring succeeding winters will add to the glacial mass and encourage additional glacial formation and movement, but there are simply no answers to the critical question. What factors or ingredients are known to initiate cooling and maintain the temperature low enough for long enough so further cooling, higher snowfalls, thicker ice on ponds and more winter weather became the norm.

Similarly not understood are the ingredients that sustain glacial growth for millions of years. Nor do we really know what causes intermittent, warmer interglacial periods, the eventual melting and retreat of the ice and he remarkable return to normal climates after so many very cold years.

Volcanoes, Earth's orbit, reforestation, the presence of large amounts of various chemicals such as sulfur dioxide, carbon dioxide and methane in the atmosphere, the motion of Earth's tectonic plates, changes in the speed and direction of ocean currents caused by changes in water temperature and salinity, meteor explosions. Chunks of the sun breaking off heading to Earth and causing additional warming. These are all possible causes.

The last glacial maximum was the period in Earth's history when ice sheets extended their greatest distance. This occurred about 26,500 years ago. Vast ice sheets covered much of North America, northern Europe and Asia. A little known consequence of glaciation was the profound effect these massive ice sheets had on the earth and its climate including drought, deforestation and a dramatic drop in sea level. That portion of the earth's surface not covered by ice was dramatically drier and cooler. In North America the ice covered essentially all of Canada, extended southward to the Missouri and Ohio Rivers, and eastward to New York City. It's hard to imagine, and hard to understand, but it did happen. And they did disappear.

Taking the Earth's Temperature – Detectives and Observers

The Earth's climate has changed many times during the 4.6 billion years of its existence. Getting accurate data with which to study climate and accurately assess climate change over those billions of years (and more specifically today) is indeed a challenge. Instrumentation of the type and accuracy we use today to measure meteorological variables like temperature and precipitation have been used for literally only the blink of an eye in Earth's climate history. One way to characterize past climates is to group them into timeframes based on the methodology used to construct or, in most cases, reconstruct climate conditions. These timeframes overlap. As time goes on, our understanding of climate increases, but only ever so slowly.

Timeframe	Climate Characterized By	Length of Timeframe
Palaeoclimatological	So-called "proxy" data obtained from geological records like tree rings, lake bottom sediment and ice cores	As far back as proxy data can be found
Historical Records	Accounts of droughts, floods, especially hot summers and cold winters, storms, timing of harvests and the like passed down from generation to generation through stories and later in some form of written documentation	The span of recorded human history, approximately 5,000 years
The Age of Instrumentation	Direct measurement of weather data using calibrated, quantitative, reliable, repeatable, robust scientific instruments	At most 150 years (from the mid 19th century to present day)

When it comes to taking the Earth's temperature, two very important groups of players come to the forefront. Those who use clues from the physical remains of various animate and inanimate objects to assess and define climate – the ***Detectives***.

Those who, before the advent of the Age of Instrumentation, measured weather parameters using rudimentary tools, noted and described significant weather events, and documented that information – the ***Observers***.

The Detectives - Paleoclimatology

Paleoclimatology is the study of climate throughout the Earth's history using information locked in geological and other "vaults". Valuable information may hide in the remains of living organisms like diatoms, coral reefs, and in lake beds, exposed sedimentary rock formations, ocean bottoms, glaciers and tree rings. With such sources, paleoclimatologists can glean temperature, humidity and precipitation information from long ago, and develop climate proxies. Such proxy information covers different periods of the Earth's history, for different lengths of time, and with different levels of resolution and precision.

For example, tree ring proxy information is of value for at most several thousand years; ice core data for hundreds of thousands of years, and fossils for millions of years. So, tree rings can't help solve climate questions during the age of the dinosaurs, but ice cores can. And all this data is subject to questions of resolution. For instance, fallen trees generally remain relatively undisturbed and annual tree ring growth can usually be considered to have good to excellent resolution. On the other hand, lake bed sediment has probably been subjected to the forces of wind, waves, ice formation and the like. This discombobulation leads to questions about the accuracy of the resulting paleoclimatic data.

Of course, paleoclimatic data can never compare to what is obtained from today's sophisticated weather instruments. Nonetheless, <u>when all available information for a given location is combined and assessed, *paleoclimatic proxies can provide a remarkably accurate picture of climate.*</u>

Types of Paleoclimate Proxies

Below is a table that summarizes, compares and contrasts the more common Paleoclimate proxies used for assessing and estimating past climates before the presence of human history and the introduction of first rudimentary and later sophisticated sensors and other instruments for measuring temperature and other meteorological parameters.

Proxy	Tells Us	How		
Tree Rings	General climate and especially precipitation in a given growing season or year	Annual tree rings vary in thickness, color, chemical composition	Good – many trees end up in stable, undisturbed environments	Perhaps as long as several thousand years
Lake and Seafloor Sediment	A myriad of climate information from an assortment of deposits	Deposits include sand, rock fragments, eroded soil, wind-borne volcanic dust and pollen, skeletal remains	Variable — depends on the parameter being assessed. For instance, high resolution found in lakes that contain little or no oxygen	Hundreds of millions of years
90 million-year-old Sedimentary Rock	Link between variations in the orbits of the planets in our solar system and climate change	Two planets coming in relative proximity to each other exert a slight tug that alters their orbits, affecting the amount of solar radiation Earth receives from the Sun	Improving — relatively new area of study centered at present in Big Bend, Texas	Geologic Time Scale

Ice Cores	Annual snowfall, concentration of greenhouse gases, timing of volcanic eruptions and dust storms, solar activity, ocean water temperatures	Annual ice thickness, presence and composition of air pockets and aerosols, concentration of beryllium-10, ratio of oxygen isotopes	Good — in some cases year-to-year comparisons are possible	Hundreds of thousands of years
Coral Reefs	Changes in sea level, turbidity, temperature, and El Nino cycles	Chemical composition and layering of reefs	Seasonal	Several hundred years
Packrat Middens	Evolution of plant communities and timing for extinction of animals	Analyzing articles like clumps of vegetation, insects, remains of vertebrates	Good	Upwards of 20,000 years

A little more on ice cores, ice cores drilled to-date in Antarctica and Greenland, the largest ice sheets in the world, provide climate information going back 800,000 years. The deepest ice core drilled by a U.S. team was removed from the West Antarctic Ice Sheet. It was a 4.8" cylinder that reached a depth of 11,171 feet. It provides detailed information and environmental conditions that have occurred during the past 68,000 years. Gleaned from this compacted frozen snow was information on greenhouse gas concentrations, wind patterns and ocean temperatures.

The Observers - Historical Records

Back before man could accurately measure temperature, if in fact he could measure it at all, weather was occurring all around him. Some of this weather had to be extraordinary, the kind you don't forget. Imagine, in the absence of any understanding, and with no 5-day forecast the impact of a landfalling Category 4 hurricane, an F4 tornado a half-mile wide moving at 50 mph, or floods caused by weeks of heavy rain. Such events would undoubtedly become lore, never to be forgotten.

Indeed, these weather stories would be passed down from generation to generation, generally by word of mouth (perhaps using sign language), not to be forgotten, to be perpetuated. Of course, it goes without saying that the accuracy of such handed down stories is questionable—the year(s) it occurred, how much rain fell, how many days in a row it didn't rain. In the absence of legible writing, storytelling would be the only way to "document" the weather. Interestingly enough, modern man (that means us, you and me!) also likes to pass down stories of "unforgettable" weather events hurricanes like Andrew, Camille, Katrina; snowstorms, derechos, tornadoes.

And in the midst of such story-telling, the story itself sometimes gets bigger than it actually was—higher winds, deeper snow; we all remember the stories of walking to school with snow up to our waists, every winter. The reality is, when looking at climate data, snow patterns really haven't changed much year to-year and especially decade-to-decade. But our orientation, our perception, indeed even our need to make it all sound a little windier or snowier than it was, all contribute to less than accurate historical memories.

Despite all this, and although the details of early man's historical weather observations may be sketchy and the absolute accuracy questionable, the *observers still play(ed) an important role in helping climatologists to better understand climate.* Each story adds a little more understanding. Taken together, these historical records are another valuable tool to help fill in gaps or verify previous assumptions. Slowly but surely, first through drawings and later with the written word, climate-oriented information was documented.

Early man experienced many of the same extraordinary geological and meteorological events that occur today. Rudimentary cave paintings and other pictographs allowed early man to "document" especially notable phenomena. Among the earliest examples of such archiving are paintings found in the Chauvet caves in southern France that appear to describe a massive volcanic eruption. These paintings are 37,000 years old. 600-year-old murals found in a cave in central Turkey also appear to be maps showing the location of nearby volcanoes. Such depictions are important to climatologists because it is well known that sulfurous gases released during and after volcanic eruptions mix with water vapor to form acids that reflect solar radiation and lead to temporary cooling of the Earth.

With the advent of the written word, detailed information about weather and climate could more easily be documented. For instance, Pliny the Younger's detailed description of the eruption of Mount Vesuvius in 79 A.D. has provided invaluable information regarding an event that would otherwise not be available to climatologists and others.

<u>Assessing a Slice of Civilization</u>

A visit to Mesa Verde National Park in southwestern Colorado provides an intriguing story of the Anasazi Indians who lived in cliff dwellings and grew corn, maize, squash and beans on the roofs (the Mesas) of their homes. Their civilization thrived during the 1100s and 1200s. Successfully growing these crops required a minimum amount of rainfall and a minimum number of frost-free days. From the start they were close to these minimums. Over the course of decades, the climate in southwest Colorado became dryer and colder. The crops failed; the Anasazi moved on. Although there is probably more than one reason for the demise of the Anasazi, we know that their abandonment of Mesa Verde coincided with a period starting around 1250 that was characterized by erratic rainfall and periodic drought, and that starting around 1270 the weather was characterized by cooler temperatures and a long term drought.

Brother Joseph Dietrich—An Early Observer

Four hundred years later, at a 16th century monastery in Einsiedelin, Switzerland, Brother Josef Dietrich began to document the day-to-day weather he was experiencing. The result was thirteen bound diaries with detailed meteorological observations dating from 1671-1704. These observations pre-dated by hundreds of years the emergence of calibrated, repeatable instruments like thermometers. Until just recently, compilations of weather observations like Brother Dietrich's were viewed as curiosities, stories. But a closer look and deeper analysis of this and other seemingly "interesting but inaccurate" weather stories has found that in fact this action is very useful, especially if it spans many years. The records provide a context for assessing climate change. The information fills in gaps in our weather history, provides data and trends for periods of time largely void of useful data.

Especially remarkable events such as significant snowfalls, extremely cold temperatures or landfalling hurricanes, potentially otherwise unknown, become part of our weather history. Even without sophisticated instrumentation at their disposal, these weather observers created databases that are perhaps more accurate over say a 20-40 year period than proxy data from tree rings would provide. Here is an entry in Brother Dietrich's weather diary from January 11, 1684: "It was so frightfully cold that all of the communion wine froze" (apparently a phenomenon never experienced before). And then "but two days later it got even worse". Assessment of Dietrich's tireless work showed that the period from 1671-1704 was one characterized by significant extremes. There were both very cold winters and very mild winters, and very dry summers and very wet summers. So perhaps every severe weather event or out-of-whack season is not as unprecedented as we are led to believe. Extremes were occurring 500 years ago!

Thomas Jefferson – Weather Observer Extraordinaire

In terms of the number of locations at which weather was observed, the consistency and continuity of the data he captured, and the types and variety of the observations he made, probably no one added more to our written knowledge of meteorology and climate prior to the advent of the

Age of Instrumentation than Thomas Jefferson, principal author of the Declaration of Independence and third President of the United States of America. He was determined, indeed obsessed, to understand the climate of this new country.

Jefferson made weather observations at Williamsburg, Virginia, at his home at Monticello in central Virginia, and at some European locations to which he travelled. For 50 years! It was at Monticello that he made by far his biggest contribution. He rose as early as 3 or 4 a.m. to take a temperature reading after concluding that this was typically the coldest time of the day. And he took another reading at 4 p.m., the time of the day he concluded was most likely to yield the high temperature of the day. Although we now know that this is not true every day, it was the consistency of this approach that was most important, providing data that could be effectively compared across years. And of course he collected, measured and documented daily precipitation. His attempts to collect wind speed and humidity data were hampered by the absence of suitable instrumentation for those purposes. Nonetheless Jefferson produced, day after day, a daily weather story that would be the envy of any amateur and probably many professional weather observers. Jefferson realized that it would take far more than minimum temperature, maximum temperature and total precipitation to characterize any day meteorologically. Jefferson's reports provided not only temperature and precipitation data, but also a shorthand description of the observed weather conditions or any remarkable changes in those conditions twice a day.

- So, for example, by using the letter "a" for "after" and using the first letter of each weather parameter, Jefferson could denote "f a r" to describe "Fair After Rain".
- He also had a "Miscellaneous" column that allowed him great latitude in documenting unusual, seasonal or extraordinary events, like the timing of the emergence of the first cherry blossom, or the first measurable snowfall.

Here is what Jefferson's Daily Weather Log might look like

Date   Morning
Temp   Morning
Weather   Afternoon
Temp   Afternoon
Weather   Rain (")   Miscellany

Date	Morning Temp	Morning Weather	Afternoon Temp	Afternoon Weather	Rain (")	Miscellany
Oct 24	35	F A R	49	F	0.14	
Oct 25	29	F	45	C A F		Frost a.m.
Oct 26	38	R	52	R	1.03	Heavy Rain

Thomas Jefferson was, <u>as a weather observer, way ahead of his time</u>. Today, over 11,000 cooperative weather observers regularly collect and provide to the National Weather Service daily meteorological data to supplement that collected at the 122 Local Weather Forecasting Offices. But today's observers, I think would agree, have it just a bit easier than Thomas. Even a modern, middle of the road personal weather station automatically provides a daily report of maximum and minimum temperature, mean temperature, total rainfall, highest hourly rainfall, etc., etc., etc.

Taking the Earth's Temperature – the Data Handlers

Any study of the Earth's climate takes us back millions of years.

For most of this extraordinary timescale, it was not possible to directly measure temperature, and climatologists had to use "proxies", data and information obtained from or the last 5000 years or so, Homo sapiens provided us with generally non-scientific stories, sketches and written records of the weather they observed and experienced, and then the thermometer arrived. Suddenly it was possible to measure the temperature of many substances and, from a meteorological standpoint, to collect and analyze actual atmospheric temperature data.

Whether we are checking to see if a grandchild has a fever, feeling the water with a hand before stepping into the shower, trying to decide whether or not to wear a sweater or, from a more global perspective, comparing how warm it was compared to a year ago, we are all trying to determine the same thing— The Temperature. Here are some examples of definitions of temperature:

- A measure of the average kinetic energy of the particles in a sample of matter, expressed in terms of units of degrees designated on a standard scale
- The degree or intensity of heat in a substance, especially as expressed on a comparative scale and shown by a thermometer or perceived by touch
- A numerical representation of hot or cold compared to one or more baselines, typically the points at which water boils and freezes

The Development of the Thermometer

The precursor of the modern thermometer was the thermoscope, which was essentially a thermometer without a scale. In 1593, Galileo Galilei invented a rudimentary thermoscope using a glass bulb containing various fluids that expanded or contracted with changes in temperature. As with all previous inventions, Galilei's design was susceptible to changes in atmospheric pressure. In 1742 Swede Anders Celsius introduced the Celsius Scale, which divided the temperatures between water's freezing and boiling points into 100 units. As a result, individual units on the Celsius scale are often referred to as "degrees centigrade". In 1948 the Celsius Scale was adopted as the preferred temperature scale by an international conference on weights and measures. It is used in virtually every county in the world except the United States, Thomas Jefferson, the third President of the United States and a renowned weather observer, bought more than a dozen brass, mercury-filled glass thermometers built in London in 1788 by John Jones. Several years later he bought a similar instrument built in Philadelphia for the sum of three pounds, 15 schillings, which would be more than $500 in today's currency.

The advent of the thermometer made it possible to quickly assemble the data needed to compare and assess climate. No need to look at ice cores or tree rings or lake bottoms; no need to sit on grandpa's lap as he recounted (qualitatively of course) how cold and snowy it was in North Jersey when he walked to school in unimaginable weather conditions. But early thermometers were still not considered a source of data accurate enough for scientific studies of climate. Could the thermometer provide "reproducible" results, was it durable enough to withstand the weather conditions at its location, could it be calibrated. From a climatological standpoint, 1850 is generally accepted as the start of the "Instrumented Records Era" though it's quite clear substantially improved sensors did not become available for decades after that.

Calculating the Earth's Temperature

To believe that scientists can calculate an annual, accurate temperature for the entire Earth's surface to the nearest hundredth of a degree that is unequivocally accepted by the technical community may prove difficult for many people to accept.

After all, there are vast expanses of the ocean's surface and until recently only very rudimentary means to measure its temperature. There are places on land in locations like Siberia that few humans have ever set foot on, let alone had an Automated Surface Observing Station (ASOS) to measure daily temperatures. Every day the difference between the highest and lowest temperatures recorded on the Earth's surface is more than 100°F. Plus there are dozens of questions about individual sensors—how old or new are they (both lead to potential inaccuracy issues), where are the thermometers located, is data from satellites helping or not?

Since the late 1800s, surface-based thermometers have been the primary data collection tool for climate studies. Although these thermometers are measuring so-called "land temperatures", the sensors are in fact in contact with the air above the land site, not with the surface of the Earth itself. So, formally, these are actually "near surface temperatures". Thermometers are typically positioned 4-6 feet above grade. Thermometers should be placed inside a "Stevenson Screen", which is a white plastic or wood housing that comes in a myriad of designs. It provides good ventilation, and keeps rain and snow off the instrument. Most importantly, the "Stevenson" shields the thermometer from the direct rays of the sun, ensuring all readings will be standardized, that is, "in the shade"

ASOS's (Automated Surface Observing Stations) are placed in remote sensing locations without the need for human participation in the measurement process. While the situation has improved, electronic automated rainfall accumulation techniques still have limitations and are not as accurate as one might think.

A team of meteorologists was assembled by NOAA to examine and assess thermometers used as official sensors at a number of the 122 National Weather Service Local Weather Forecasting Office locations. These instruments are sophisticated sensors, the best available. The assessment confirmed that it is not uncommon for the air above the thermometer location to be affected by local heat sources. Among others, the thermometers were found to be located near the exhaust fans of air conditioning units, surrounded by heat-absorbing asphalt parking lots, on extraordinarily hot rooftop, near sidewalks and buildings that absorb and retain heat, and near wastewater treatment plants whose chemical reactions expel heat. Over half of the temperature sensors did not meet the NWS's own siting requirements.

Urban creep occurs when thermometers once in a relatively cooler rural locations are "suddenly" registering warmer readings caused impinging roads, buildings and parking lots, unrelated to today's higher global temperature and based stations positioned near the water may be subject each day to cooling onshore winds, exposing the sensor for portions of each day to temperatures not representative of the thermometers land-based position.

Ocean surface temperatures have only recently been measured by modern techniques like ship intake temperature monitoring, static and drifting buoys and infrared satellite measurements. Prior to that sea surface temperatures were measured quite primitively by sticking a thermometer into ocean water collected in wooden or canvas buckets

Satellites – Wide Coverage, Filling Gaps, Questions About Accuracy

Thermometers generally provide accurate temperature data from single fixed locations. In some cases, those sensors have been recording data for decades, providing a nearly unbroken string of valuable temperature data over a long enough period to identify trends of warmer, colder or stable temperature. Satellites, on the other hand, provide much greater geographic coverage. With the exception of the poles, satellite readings can provide nearly complete coverage of the Earth's surface. Satellites are increasingly being used to validate and fill in the holes of the Earth's temperatures where thermometers can't be deployed. But questions continue to arise about the accuracy of temperature data collected by satellites when clouds and rain are present. Similarly, questions are being asked about the ability of satellites to distinguish between unfrozen surface water, and ice and snow. In addition, satellites have only been providing data to climatologists' calculations of the Earth's annual temperature since 1969. When it comes to determining the Earth's average temperature, satellites are relative newcomers. Finding the best role for satellites is still a work in progress.

Radiosondes are weather instrument packets carried into the atmosphere by balloons. They send back continuous temperature data as they ascend through the atmosphere. The radiosonde is the small white box at the bottom of the screen that contains the data-gathering instrumentation

The All Important Anomaly

Climatologists start with all the available data that has been accumulated in the four major datasets to help calculate an average global temperature. The temperature at each land and ocean station is compared on a daily basis to what is "normal" for that location and time (date). This is typically the 30-year period used globally by climatologists. The differences, warmer or colder, are called "anomalies." This is a very important part of the process—and perhaps a bit surprising. Just to reiterate, climate change is based on the difference in temperatures from one year to the next recorded at the thousands of temperature sensing stations throughout the world, and not on absolute temperatures.

Daily anomalies are averaged together over entire months, which in turn are accumulated and averaged to provide season-to-season, year-to-year anomalies lines. After determining the annual temperature anomalies for each land and ocean station, the Earth's surface is divided into grid boxes. The average temperature/anomaly for each box is determined by combining all available data for that little spot on the Earth's surface. The smaller the grid boxes, the more accurate the average temperature will be. At times the determination of an annual average global temperature can sound quite simple, requiring nothing more than rudimentary arithmetic to calculate the anomalies and a straight edge to draw the grid lines.

Can We Really Trust All the Data That Determines the Average?

- We know that many thermometers are reading one or more degrees warmer or colder than the actual ambient conditions in which they are placed.
- We know that climatologists used thermometers over a hundred years ago that were most probably reporting inaccurate temperatures.
- We know that all thermometers periodically require recalibration but aren't receiving them.
- We know that newer temperature gathering tools like satellites and radiosondes have been in use for only a matter of decades and that the earliest data gathered by these techniques may not be as accurate as initially thought.
- We know that many locations on this huge Earth of ours have never had their temperature measured by an actual thermometer.

So, is it really possible, with all those shortcomings and obstacles, to <u>determine</u> an accurate annual global temperature?

The simple answer is, <u>"YES"</u>, we can calculate with some reasonable confidence a global earth average temperature by combining and integrating readings from individual temperature sensors.

And the reasons?

1. The "Anomalies"

2. Thousands upon thousands upon thousands of individual temperature sensors in a multitude of locations

By using anomalies, temperatures are based on the changes since last year rather than actual absolute temperatures. So a thermometer that is poorly calibrated, or inaccurate, or placed near asphalt or an air conditioner outlet, registers the same "bias" year after year. The exact temperature may not be correct, but the year-to-year change is.

Thousands Upon Thousands of Data Points

And, although any one thermometer is far from accurate, the inclusion of so many individual data points, added together, provides a more than acceptable means for detecting and tracking changes in Earth's temperature over time. Previous articles about climate data have described the "amazing consistency of the 30-year average", meaning 30 years is sufficient to "smooth out" the extremes even when one year, or even a day here or there, is extraordinarily hot or cold. Calculating an average temperature for the earth benefits from the same statistical reality. Outliers, poor calibration, human error and any number of other temperature measurement snafus can be swallowed up and accommodated by the average of the sheer volume of data collected.

It seems to work. At least the scientists believe it does.

Next are two climate oriented stories, the first a short overview of fall foliage and the impact various weather phenomena, such as temperature and rain and wind have on the timing, colors and length of fall foliage.

The second article is a story, the subject of which is quite simply "The Dust Bowl". But there is so much to this story. The article will review the homesteaders, farming techniques, even the role of WWI, and much more.

Fall Foliage and the Weather

There is really no way to measure the comparative beauty of Mother Nature's annual foliage displays. Absolutely brilliant reds, oranges and yellows, lasting for weeks, refusing to give into the bleak blacks, greys and whites of the upcoming winter.

So how is it that this wonderful beauty comes our way? And what acts to intensify and prolong this beauty or, from the opposite perspective, blunt and shorten it?

Shorter days (less sunlight) and cooler temperatures trigger chemical changes in leaves.

Phytochrome is the light-sensing mechanism that recognizes the shorter days. Chlorophyll, which is the leaf pigment responsible for the green color of most leaves, breaks down quickly amidst the changes in daylight and temperature, opening the door for yellow and orange pigments to form. Yellow is by far the most prominent autumn leaf color. Fewer leaves turn red because anthocyanin, responsible for red pigment, is not found in nearly as many trees as the carotene and xanthophylls that produce oranges and yellows. Red is the most sought-after leaf color and tracts of land with significant red components are often considered the most beautiful.

The only certainty associated with fall foliage is that days will get shorter, and temperatures will trend cooler, as autumn gets underway. Enter the weather. Many meteorological factors affect leaf foliage. The four with the greatest influence are temperature, moisture, sunlight and wind.

Although the variables that affect leaf color are understood, no one yet knows how much or how little each contributes to the onset, duration, variation and brilliance of autumn color. As it turns out, it's much easier to explain "why?" after the colors have formed and the leaves are gone than it is to predict it before autumn gets underway.

- Temperature – an early frost can severely affect fall foliage. Below freezing temperatures can kill leaves outright or accelerate leaf droppage. Provided temperatures stay above freezing, cool temperatures tend to favor anthocyanin production, enhancing red colors.
- Moisture – mild drought favors red leaf color. A strong late summer drought may delay the onset of color for several weeks. Rainy days occurring near peak coloration tend to decrease color intensity.

- Light – more sunshine favors red coloration, especially for those shrubs receiving direct sunshine. Cloudy conditions tend to suppress color brilliance
- Wind – the passage of a vigorous cold front with gusty winds, or a prolonged period of high winds associated with for instance a nor'easter, can strip leaves off branches, limiting the duration of their colorful show.

Since temperatures, the amount of cloud cover, rainfall (both timing and intensity) and the presence or absence of wind vary from year to year, it is certain that no two autumns are ever going to be the same when it comes to fall foliage. Why was the foliage so spectacular in Northern Virginia this autumn of 2022? The mild drought? The number of warm sunny days? The general absence of windy conditions? The cool nights and lack of hot days? Maybe. The changes in weather and daylight as autumn presses on also trigger a hormone that releases a chemical message to each leaf that it is time to prepare for winter. Over the next few weeks, abscission cells form a bumpy line at the place where the leaf stem meets the branch. And slowly, but surely, the leaf is "pushed" away from the tree branch.

It's time to knock the rust off the rakes and blowers and collection bags. Those once colorful leaves will soon turn brown and gradually become part of the next layer of topsoil. And so will this process repeat again and again.

The Extraordinary U.S. Weather Decade of the 1930s

The Intro

Weather tends to stick close to climatic norms. There are always exceptionally hot summers, or particularly severe winters, or a string of very wet months to catch the attention of climatologists. But over the long haul daily, monthly, and annual temperatures, rainfall and snowfall generally are in line with 30-year climate averages, 30 years having been deemed by the World Meteorological Organization way back in 1934 to be a sufficiently long period of time to absorb exceptional weather events and maintain a "smooth climatic curve".

Then there were the 1930s—one could argue the most extraordinary weather decade in recorded American history. No decade had before, and no decade has since, come close to setting the number of record individual state high and low temperatures, suffering the lack of rainfall, or being overwhelmed by dust storms the scope of which had never (as far as we can tell) occurred. Additionally, the decade produced several exceptional weather events, many tropical storms, including several particularly intense hurricanes, that add further credibility to the claim that this decade of the 1930s was indeed the most extraordinary U.S. meteorological event.

I have always been fascinated by the immense meteorological stories of the 1930s, especially when interwoven with the unstoppable optimism and sheer grit of those homesteaders and other settlers who came to the Plains to carve out a living only to be swallowed up by the weather that for some reason descended upon them in the 1930s, a generally quiet time between the two world wars when America was going through its westward migration

How and why this crazy, exceptional weather happened has never, to my knowledge, been determined. Just a blip in whatever it is that causes the weather? If nothing else, it shows how fickle and unpredictable the weather can be. To some extent, the 1950s and 1960s were a little bit like that. Worries about global cooling surfaced. And there have been quite extraordinary snowfalls in relatively recent decades that rewrote the snowfall records of many U.S. East Coast cities, and so on. There is plenty to tell in this story of the extraordinary weather decade of the 1930s. Here are the players, animate and inanimate (you'll understand as you read on):

The Heat

- The decade of the 1930s was profoundly hot, affecting virtually the entire continental United States.
    - o The summers of 1930, 1934 and 1936 are generally regarded as the hottest years of the 1930s.
- 1934 featured a heat wave in which many parts of the country suffered through close to thirty straight days of temperatures over 100F.

- 1936 was characterized by shorter but hotter periods of extreme weather that produced temperatures far above those anyone had experienced at any time in the past or, for that matter, since.
- The map below that shows the average temperatures for each state in 1936.

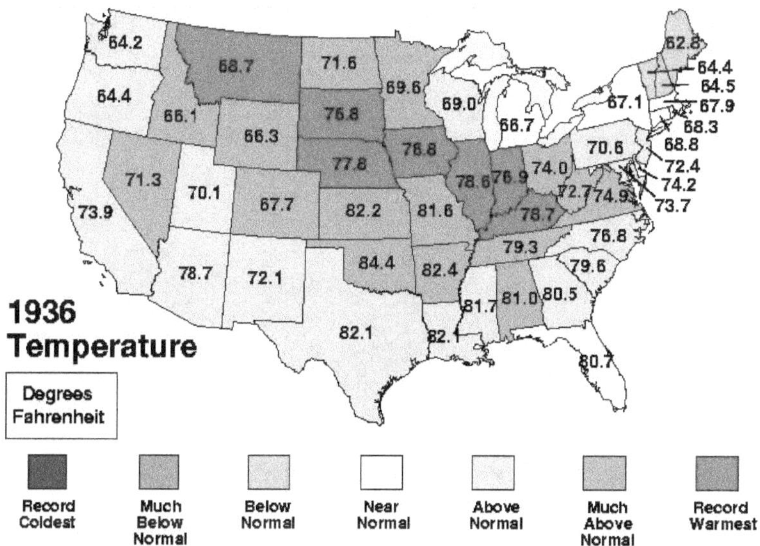

- There was nowhere and no way to escape ongoing day-and-night heat of this magnitude, especially in the middle two-thirds of the country.
- Many all-time high temperature records were set across North America in 1936.
- Thirteen states had their highest recorded temperatures ever in 1936, ranging from 109F in Cumberland, Maryland to 122F in Altus, Oklahoma.
    - o Arkansas, Indiana, Kansas, Louisiana, Maryland, Michigan, Nebraska, New Jersey, North Dakota, Oklahoma, Pennsylvania, West Virginia, and Wisconsin
- Remarkably, these record highs still stand today, 90+ years later.
- Ten other states established record high temperatures in the decade of the 1930s.
    - o These states were Delaware, Kentucky, Mississippi, and Tennessee in 1930; Florida in 1931; Idaho, Iowa, and Ohio in 1934; Montana in 1937 and Oregon in 1939
- Most of these records, too, still stand today.

- All told, 23 states, almost half of America's lower forty-eight states, had their highest recorded temperature in the decade of the 1930s.
- The chart directly below contains some minor inaccuracies but nonetheless provides an excellent visual representation of the extraordinary impact the hot weather of the 1930s had in establishing individual state record high temperatures.

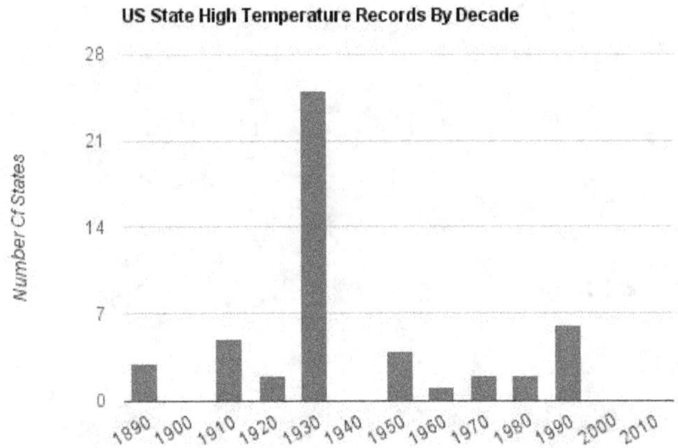

Could one decade really account for nearly half of all the states' highest temperatures on record? High minimum temperature records were set in many locations, including "lows" of 91F at Lincoln, NB and 94F at Atchison, KS the morning of July 25, 1936. The Heat Wave of 1936 caused catastrophic human suffering and death. Air conditioning was still a dream. Apartments into which people crowded quickly became more like ovens. More than 5,000 people died.

The Cold

The 1930s were also exceptionally cold in North America. The 1936 North American Cold Wave ranks among the most intense ever recorded. February 1936 was the coldest February on record in the contiguous forty-eight states, surpassing the cold of February 1889. It was the coldest month ever in Nebraska, and North and South Dakota. The meteorological winter that ran from December 1935 through March 1936 was the coldest on record in Iowa, Minnesota, and Dakota.

Nine states set record low temperatures in the 1930s—California, Michigan, Nevada, North Dakota, Oregon, South Dakota, Texas, Vermont, and Wyoming. Three states had both their record low and record high temperatures in the decade of the 1930s. These were Michigan, North Dakota, and Oregon. North Dakota had both its record high (121F) and its record low (-60F) temperature in the same year—what else?—1936.

It took only 141 days for North Dakota to go 181 degrees from a record low to a record high.

The Dry

The 1930s not only set records for cold and heat, but they were also extraordinarily dry as well. The Drought Monitor, a resource provided by the Dept of Agriculture in association with the University of Nebraska and NOAA showed that exceptional drought occurred in 1934 in all of Colorado, Wyoming, North Dakota, South Dakota, Nebraska, Kansas, Iowa and Minnesota, and in parts of fourteen other states.

The droughts were at their most severe in 1934, 1936 and 1939-1940. Some locations suffered through drought for as long as eight years. 1933 was the driest summer ever recorded in the Oklahoma panhandle. Rarely if ever in recorded history has a drought of such scope and severity been experienced in the Plains states of North America. A detailed and exhaustive study of drought in North America found that the drought of 1934 was the driest and most widespread of any drought experienced in the last 1,000 years. The next driest year occurred in 1580. The drought year of 1938 affected almost three-quarters of North America

The Grasslands

For centuries grasslands had flourished on the American Plains. These grasses had protected the soil from rain, snow and wind. The climate of the Plains is "semi-arid", meaning the annual precipitation is about 20". Of course, average rainfall over 30 years includes years of both relatively copious rain totals and those where little if any rain fell. The indigenous shortgrass prairie biome has adapted to these variations and is able to survive through the inevitable variations of wet and dry conditions.

The Homesteaders

Many who populated the U.S. Plains were homesteaders, and many of them took advantage of the Homesteaders Act of 1862 that allowed people who were willing to stake a claim for 160 acres in then largely uninhabited lands in the "West" to begin a new life in this land of endless grass. If they were able to turn their parcel of land into a successful enterprise within five years, the land was theirs. What a success homesteading turned out to be. Over four million claims were processed, encompassing ten percent of America's land mass. States (and the percentage of their respective land areas that resulted from homesteading) include Nebraska 40%, North Dakota 39%, and Oklahoma 34%.

The Farmers

The majority of those arriving at their 160 acres of grasslands had no farming experience. Many early settlers initially chose cattle ranching as their primary source of income. Grazing of grasslands led many to consider agriculture instead. Unusually wet weather in this semi-arid climate in the late 1920s encouraged farmers to continue converting lands to agricultural usage. Many of the early settlers used horse or oxen drawn plows to turn over the prairie sod.

These were soon replaced by mechanized equipment, capable of increasing the rate of sod removal by a factor of four or more. Wheat was thriving, and attracting handsome returns on investment, especially when the weather was good.

The Mistakes

Several years of above normal rain not only provided improved profitability but encouraged even more sod removal and increased wheat cultivation. In 1931, just as major meteorological changes were beginning over the Plains, a record harvest of 240 million bushels was grown. Wheat shortages brought upon by the Russian Revolution and World War I further added demand and raised prices.

Despite record harvests, the new homesteaders did not follow the basic rules for plowing and crop rotation. Grasslands that had for centuries protected the soil from rain, drought and wind had been sliced, removed, or otherwise compromised. Too busy counting the profits from wheat, farmers failed to practice and apply dryland farming methods that would have minimized soil erosion. Plowing of the virgin topsoil led to exposed soil much less resistant to erosion. New acreage translated to new plants translated to more crops and more profit—nothing could stop the greed, except a powerful drought.

The Dust

By removing grasslands with deep roots that anchored the grass to the soil, and could seek water well beneath the Earth's surface, farmers had unknowingly laid the groundwork for terrible consequences down the road. Or perhaps they knew but just couldn't resist just one more spectacular wheat harvest. The 1920s ended with another wetter than normal year, surely a good omen for next year and the coming decade. Then the rain stopped. 1930 ushered in the decade with an unusually dry summer. Four of the next seven years would suffer below normal rain. Unprotected topsoil became friable and powdery in the intense sun and high temperatures. The unanchored soil turned to dust. It was no problem for the winds of the Plains to lift and transport the soil.

The stage was set for one of the greatest meteorological disasters in America history.

The Dust Storms

At first there were just some small, swirling dust clouds, like the dust devils seen in the dry lands around Phoenix. Then larger dust clouds appeared, at first more of a curiosity but with time growing larger and more ominous. As more and more dust became airborne, visibilities vanished. Increasingly the dust storms became more frequent and more menacing. Soon clouds of unimaginable size, speed, and destruction appeared. The enormous, billowing, fast-paced clouds were called "black blizzards" and "black rollers". During such storms, the visibility could be reduced to less than three feet. Such clouds had never been experienced. A dust storm is defined by the

NWS as quite simply "a severe weather condition characterized by strong winds and dust-filled air over an extensive area". Roughly speaking a wind speed of 30 mph is required to move dirt. At 40 mph or higher, a dust storm may be formed.

A dust storm is propelled by both upper air jet streams and by surface winds. Dust storms wreaked tremendous damage on agriculture, the ecology, homes and farms and, most of all, people. In 1932 alone, Oklahoma lost an estimated 440 million tons of topsoil to dust storms. A May 1934 dust storm engulphed Chicago in over 12 million tons of dust. This same storm moved eastward into the Atlantic Ocean, coating ships with a layer of dust. So-called "Black Sunday", April 14, 1935, is calculated to have stripped an amount of topsoil from farmers' lands equivalent to more than twice the dirt dug to create the Panama Canal

The Dust Bowl

The term "Dust Bowl" was coined in 1935 by AP reporter Robert Geiger after he had experienced the storms while on assignment. At its peak, the Dust Bowl, with its epicenter in the Southern Plains, covered an area of 100,000,000 acres, about half the size of Pennsylvania.

As the Dust Bowl continued year after year, more and more land lay barren, covered by dust, incapable of supporting agriculture. By 1935 the situation had become desperate. Hunger and poverty quickly replaced the euphoria of endless harvests of just a half a dozen years earlier. Storms continued to pummel houses. Many were damaged structurally. Everybody had dust in their homes. No matter how much people tried to keep it out, the dust came in through any crack, even small ones invisible to the naked eye.

The Departures

No longer able to farm and with dust covered land unlikely to be productive for many years again. Families simply gave up and moved away. First a few, then a lot. An exodus was underway. Between 1930 and 1940 approximately 3.5 million people left the Plains states and headed west yet once again, this time mostly to California. No previous long tern meteorological event had led to such a move. American meteorologists rated the Dust Bowl the #1 weather event of the 20th century.

The Tropics

It wasn't just the Dust Bowl that led to the thesis that the decade of the 1930s was a special, unprecedented, meteorological period. How about the tropics? The five-year span from 1931-1935 was one of the busiest tropical periods dating back to 1850, generating four of the twenty-five deadliest hur-

ricanes on record. The 1933 season was the third busiest, producing 20 tropical systems and 11 hurricanes, just behind 2020 and 2005. There were
twelve landfalling hurricanes from 1931-1935. Among those was the Labor
Day Hurricane of 1935 that struck the Florida Keys with winds estimated
at 185 mph and a record-setting low barometric pressure of 26.35".

The Labor Day hurricane was the most intense and eighth deadliest to
make landfall in the Western Hemisphere. The decade of the 1930s also produced a Category 4 hurricane that came ashore in Texas; a Category 3 that
came ashore in Florida in 1933 and a Category 2 that came ashore in Texas in
1933. And then there were Marco and Laura who in late August 2020 became
only the second set of tropical systems in the Gulf of Mexico to simultaneously
directly affect the U.S. mainland. Last decade it happened in? The 1930s of
course. Later in the decade the Great New England Hurricane of 1938 formed
in the eastern Atlantic and hurried up the East Coast, crossing Long Island and
Long Island Sound before crashing into Connecticut. It was one of the most
powerful and destructive storms ever to strike Southern New England.

The Tidbit

Of course every decade has its complement of extraordinary weather
events, and the decade of the 1930s is no exception.

- In addition to the heat and cold and dust and hurricanes that made the
  1930s such a remarkable decade, here are a few more anomalies.
- On the West Coast, Eureka, CA had 26 consecutive rainy days.
- Fargo, ND endured 27 consecutive days during which the outside temperature was below zero.
- During a nationwide cold wave, East Coast rivers were frozen as far
  south as Richmond, VA.
- A major blizzard crippled much of the Northern Plains in February
  1936; Iowa was especially hard hit.
- Warmer temperatures in the spring led to significant flooding in much
  of the Midwest, Ohio Valley and Mid-Atlantic with high water marks
  that are still unbroken to this day.
- And then there is just one more little tidbit to share.

- There are only two dates in recorded history when measurable snow fell in the Valley of the Sun around Phoenix, Arizona; January 20, 1933, and January 21-22, 1937.
    - o Measurable snow has not fallen in the 80 or so years since, but both occurred in the decade of the 1930s.

The Verdict

- The tumultuous weather of the 1930s featured a significant number of records for what might be considered "traditional" weather phenomena like hurricanes, heat waves and rainfall.
- Many other decades since 1850 have had exceptional weather events, noteworthy records, and prolonged period of cold, snow or rain.
- But if you read even a small part of this article you can't avoid the fact that the decade of the 1930s contained two incredible meteorological features no other decade can claim.
- First, periods of extremely hot and to a lesser extent cold weather that led to a series of record state all-time high and low temperatures that defy any expected statistical distribution, and
- Two, dust and the devastating dust storms that continued for months.
- No other decade had them both.
- No other decade had such diversity and severity of weather.
- No, it's the 1930s.
- No doubt the most extraordinary meteorological decade in U.S. history.

The Research

A study has revealed that Atlantic Surface Sea Temperatures (SSTs) were an important factor in the Dust Bowl heatwaves, enhancing spring drought, and allowing heat to develop earlier over the central U.S. Although the role of SSTs in triggering the Dust Bowl Drought is recognized, land-cover changes like elimination of endemic grasses and poor farming techniques together amplified the drought 1920s and 1930s included widespread crop failures surpassing 60% in many counties. See Schubert, S. D., Suarez, M. J., Pegion, P. J., Koster, R. D. & Bacmeister, J. T. On the cause of the 1930s Dust Bowl. Science 303, 1855–1859 (2004)

CHAPTER 6 — GLOBAL WARMING

Recent United Nations' Intergovernmental Panel on Climate Change (IPCC) ICC reports have all the technical details and policy issues related to climate change; for this book, we'll simplify global warming down to some practical and understandable impacts that we might all observe in the coming years.

There is little doubt the world is in a period of warming; I think most of us can agree with that. Even anecdotally (forget the numbers) the many small ponds across the Northern Tier of the U.S. that used to support wintertime skating many days of the winter back in the 1950s and 1960s are nowadays unlikely to reveal even a hint of ice now.

No matter where you stand regarding the myriad of climate change issues confronting us, no matter how skeptical you are about climatologists' confidence in comparing temperature proxies from ice cores and packrat middens with the output of today's sophisticated temperature sensors, it is undeniable that warming is occurring, and quite quickly. And climatologists' forecasts consistently speak of additional warming and dire consequences. Even as we get ready to go to print, July 2023 was the warmest July in however many years they have kept track of Earth's temperature.

There is no need to analyze the numbers too closely. Just look around. Glaciers that were formed tens and thousands of years ago are retreating, rapidly. Glacier National Park doesn't have a single easily visible glacier after decades of decline. The Arctic and Antarctica ice shelves are breaking off into the ocean.

The Earth has been going through climatic cycles of heating and cooling since it was formed billions of years ago. The world was considerably warmer when dinosaurs roamed the Earth from 65 to 252 million years ago than it is today. In just the last 650,000 years there have been seven cycles of glacial advance and retreat. Even during the relatively recent Little Ice Age (1100 to

1800 A.D.), there were at least four distinct periods during which temperatures were above normal.

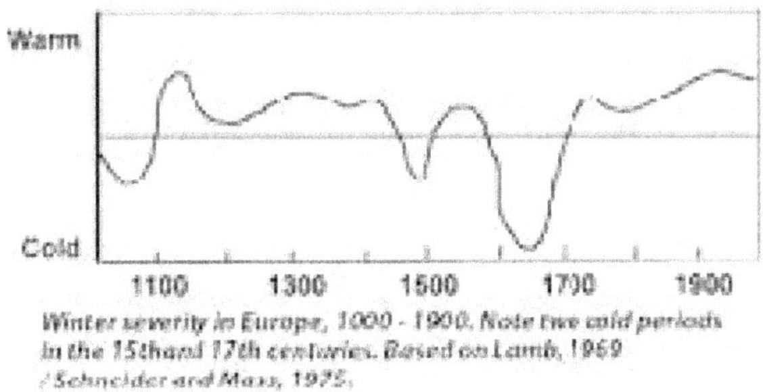

Winter severity in Europe, 1000 - 1900. Note two cold periods in the 15th and 17th centuries. Based on Lamb, 1969 (Schneider and Mass, 1975).

We still don't understand what causes these periods of colder than normal (and warmer than normal) temperatures to get started, gain momentum, slow down, stop and reverse. Many times, enormous glaciers proceeded relentlessly from the poles toward the equator, cooling the Earth as they did. Suddenly they stopped their march south and, in the presence of warmer weather, slowly retreated back to the poles.

The table below shows the average annual temperature by decade and the associated 30-year average in Washington D.C. starting in 1871.

Decade	Average Annual Temperature (F)	30-Year Average
1871-1880	55.0	
t1881-1890	54.6	
1891-1900	55.1	54.9
1901-1910	54.8	54.8
1911-1920	55.4	55.1
1921-1930	56.4	55.5
1931-1940	56.7	56.2
1941-1950	57.3	56.8
1951-1960	57.4	57.1
1961-1970	56.8	57.2
1971-1980	58.4	57.5
1981-1990	58.1	57.8
1991-2001	58.4	58.3
2001-2010	58.7	58.4
2011-2020	59.7	58.7

The overall temperature increase over the past 100 or so years is the result of, and can be described by, integrating any number of individual metrics like

- The number of summer days with low temperatures 75F or higher, or 80F or higher
- The number of daytime highs of 95F or greater
- The number of months in a year with temperatures above the 30-year mean

The number of record high minimum temperatures the extent and thickness of Arctic Sea ice has declined rapidly over the past decade. The rapid reduction in glaciers and snow cover can in part be attributed to a phenomenon known as "feedback". The more snow and ice that melt, the darker (non-ice/snow) earth is exposed. These dark areas quickly absorb more heat from the sun, accelerating the melting process. Extraordinary individual weather events need to be used carefully in justifying any position regarding climatology. With such a short time frame providing reasonably accurate data, statements like "the warmest ever" or "unprecedented" should really read something like "the coldest since recordkeeping began in 1851."

In no particular order, here are some factors that are directly or indirectly associated with global warming:

- Use of fossil fuels; sunspot activity; volcanic ash, dust and sulfur gas emissions; temperature anomalies; solar flares; thermometer calibration and accuracy issues; greenhouse gases; solar radiation management; CO2; accuracy of satellite and radiosonde temperature data; heat islands; carbon tax
- And simultaneously there is a measurement component to global climate change that is used to capture and validate outputs from various instruments and associated work processes. Examples includes high-resolution sea surface temperature analysis, thermometer calibration, and accuracy and the carbon tax

Preparing a document discussing the causes or solutions for and controversies associated with global warming would take far too much time to write and would be way too long, far too technical and much too serious to meet the guidelines I have established for this publication.

Climatologists expect warming to continue largely unabated for the foreseeable future, at least through the end of the current century. Primary concerns as the Earth continues to warm are rising ocean levels, more frequent and more severe weather events, as yet undefined but potentially significant effects on crops, other vegetation, mammals, birds and fish.

Some of the consequences being discussed about are indeed draconian.

- Entire cities inundated
- Staple crops unable to survive in hotter weather conditions
- Day-to-day weather featuring more frequent and more severe storms and droughts
- New diseases or geographic areas unaccustomed to certain diseases suddenly being confronted by them
- Nature's species moving on, or becoming extinct, unable to exist in their current surroundings in this new, warmer world
- Other challenges the human race may not be able to combat

Most climatologists see a continuing warming of our Earth with inevitable melting of remaining ice and snow, fewer snowpacks forming, more droughts, growth in deserts and so forth. On the other hand, some scientists now believe that Earth will reach a tipping point earlier rather than later. McIntyre has prepared an article entitled "Climate Tipping Points; A Personal View". Models are put on the shelf. Scientists are encouraged instead to investigate the most troublesome uncertainties like how climate and weather might or might not behave over the upcoming decades or centuries.

A recent publication warns of a global warming tipping point occurring within the next few decades associated with global warming, during and after which seawater levels would rise rapidly. The Gulf Stream would no longer exist, and severe storms would be the norm. Previous warming spells have stopped and been replaced by periods of cooler weather. Much colder weather that appeared without warning in the late 1950s and 1960s generated significant concerns about impending global cooling in less than a decade. Then, as suddenly as these exceptionally cold periods appeared, they were gone and near normal temperatures reappeared.

Many technical questions remain, many of them excellent. Unfortunately, in the midst of research aimed at acquiring a better technical understanding of global warming, a number of other activities, related but not necessarily directly relevant to, achieving the best possible technical understanding of global warming, have arisen. Global warming as a phenomenon continues to carry heavy baggage—political, monetary and ideological issues. Regardless of one's position on greenhouse gases and global warming, the current saga has, if nothing else, taught us that fewer hydrocarbons burned is better than more hydrocarbons burned. Even the most vocal disbelievers would have to admit that. And that collectively we have a responsibility to limit the creation of greenhouse gases to the extent possible.

It has been calculated that mankind has burned about 2,000 billion tons of carbon since the advent of the Industrial Revolution. This certainly sounds like a lot! And there is plenty more where that came from. The good news is that U.S. carbon dioxide emissions from all energy sources continues decline, but there are many other nations whose emissions of greenhouse gases are expanding. One country can't turn this problem around. Interestingly enough, this reduction in U.S, greenhouse gas creation has primarily been driven by market forces (less coal, more natural gas, increasingly energy efficient appliances, automobiles, LED light bulbs, etc.), not just by government regulations.

Global warming is being cited more and more as the reason for the increased frequency, severity, scope and human misery from heat waves, droughts, cold waves, hurricanes, tornadoes, thunderstorms, nor'easters, blizzards and the flooding of St. Mark's Square in Venice. Even the anomalies (record cold, major snowstorms) that seem to be in contradiction with climate change are somehow tied back in a convoluted manner to global warming.

For some, for instance, global warming is quite astonishingly being used to explain more copious snow events in the U.S. Mid-Atlantic over the past decade and the severe cold that has been experienced over the past three or four winters in many parts of Asia, Europe and North America. The three consecutive winter seasons 2013-2016 at Washington Dulles were the snowiest three such seasons since recordkeeping began in 1963.Three of the snowiest winters in New York City have occurred in the last 20 winter seasons. In reality, all of these extreme meteorological events have occurred for tens of thousands of years.

Robert Muir-Wood, in his excellent albeit extraordinarily detailed 2016 book entitled *The Cure for the Catastrophe* has compiled an impressive list of severe ("catastrophic") events caused by hurricanes, typhoons, winter storms, fires, earthquakes and volcanoes dating back to the 11th and 12th centuries.

Much of this data is from the more advanced countries of that time in Europe (e.g., Italy, Spain, Portugal) and Asia (e.g., Japan and The Philippines).

He concluded "*the extremes of the extremes of storms and floods have shown no simple trends*". For instance, up to and including the current Year of the Lord,

The most destructive European windstorm occurred in 1703.

Although the catastrophic Tacloban Philippines typhoon of 2013 was linked to global warming, worse destruction actually occurred there from typhoons in 1897 and 1912.

- The coldest winter recorded in England was 1947.
- The most extensive Central European flooding occurred in 1342.
- The worst year for hurricane casualties was 1780.

The conclusion Muir-Woods drew from his in-depth study of historically significant catastrophes is that "*for no disaster can we say scientifically that it could not have occurred without climate change*".

Another on-going theme is that global warming is enhancing severe storm intensity sufficiently that human deaths and injuries, and associated property losses, continue to increase year after year. All one has to do to feel the fear is to look back at three recent unforgettable U.S. severe storm events.

- 2005 – Hurricane Katrina – 1800+ deaths
- 2011 – Joplin Missouri Tornado – 158 deaths
- 2012 – Superstorm Sandy – $75 billion, 223 deaths

Let's take a look at the data; not just for these three events, but for "all of them". A team of experts spearheaded by Northern Illinois University, reviewed severe weather that has affected the United States since the 1850s. They concentrated on hurricanes (with data back to 1851), tornadoes, and losses from both. From this data they were able to ascertain probabilities and trends associated with these storms. Among the very interesting data they collected was, a major (Category 3+) hurricane makes landfall on the U.S. approximately once every other year (although this has not been the case since 2005). 85% of all major hurricanes have made landfall in just three states—Texas, Louisiana and Florida. Approximately 1% of all thunderstorms produce a tornado. About 1200 tornados occur in the U.S. each year, roughly 20% of which are rated as significant and 1% as violent. With some exceptions these statistics have remained remarkably consistent when looked at over decades, not just year-to-year.

So what has changed? What's different now than in 2000 or 1980 or 1940 that has led to this purported increase in human and property losses. It's not the weather. Well, it's not the weather! <u>It's us!</u>

<u>More people are living on a greater percentage of our land mass.</u>

Cities have grown in both area and population, some quite spectacularly. More and more houses, townhouses, condos and apartments are crowded along the shores of the Gulf of Mexico and the Atlantic Ocean. And this property is increasing in value, especially along ocean shorelines, where space for building is by definition limited, and the laws of supply and demand are constantly in the works. And these trends are continuing. The authors used some creative concepts and illustrations to show the differences in human and property asset losses for explicit scenarios.

For instance, the path of the infamous Category 5 Moore, Oklahoma tornado of May 20, 2013 (wind speed greater than 157 mph) was superimposed over the city of Atlanta. Had the tornado struck in 1940, the model predicts about 35 homes would have been lost. Today, such a scenario might affect close to 10,000 homes. Continued sprawl, occurring virtually non-stop, could cause the total number of homes that would be lost in Atlanta to increase to over 22,500 by 2100 using this scenario. And similar dramatic increases in loss of life and property can be shown when models are developed that address potential hurricane losses in Miami, Houston and New York City. The authors refer to the phenomenon of increased meteorologically induced losses from increased societal growth as the Bull's-Eye Effect.

Human population and infrastructure growth, not higher winds, deeper snow or more frequent storms are reasons for increased losses from severe weather. But that is not the necessarily the whole story; yes, there is evidence of humankind's role, and more and more areas are being populated now than in the 1950s, but population growth into new areas doesn't explain everything,

One thing is for sure when it comes to global warming—we have plenty of time to react. Even if the projected catastrophic results all come true, the direct effects from a worst case global warming scenario will not have tangible, life-altering consequences that need to be addressed immediately. Decades, centuries, maybe more, but not next year. Step back for a minute to 2016—for years, perhaps decades, we had been led to believe that the ongoing drought in California was among the worst "ever", and that it would continue to strengthen, driven by global warming, imperiling that state's ability to provide sufficient water for domestic and agricultural needs.

Then, in just 4 months over the winter of 2016-2107, the drought was gone! Incredible amounts of rain fell, and at higher elevations snow was so deep chair lifts were buried and skiing impossible. Another California drought cycle has just occurred. Withering drought, the likes of which have "never" been experienced were suddenly neutralized in the winter and spring of 2022-2023 when incredible moisture in the form of both soaking rains and prodigious snowfall abruptly end the drought. A look at the Drought Monitor in mid-April shows the driest conditions in the U.S. in Kansas, not California.

<u>What Are You Going to Do?</u>

Let's say, just for the point of discussion, that in fact global warming is real and temperatures around the globe are going to continue to rise. And that the burning of fossil fuels is in fact a major contributor, maybe even the primary cause. And let's assume there are tangible/quantifiable things you—yes, you personally—can do to stem the continued consumption of fossil fuels and the relentless warming. What are you going to do? What do you want to do? What should you do? What can you do? Before we go too far down the path of proactively minimizing greenhouse gas emissions, it will be instructive and useful, so we are all on the same page, to define, describe and clarify the concepts of "power", "energy" and "carbon footprint". A consistent understanding of these important concepts will help us address the actions we can take to tackle global warming.

Both nature and humans generate greenhouse gases.

Human-generated sources of carbon dioxide are much smaller than emissions from nature.

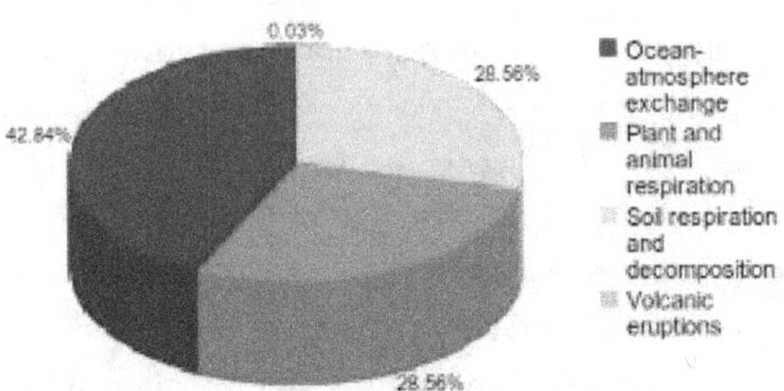

Almost half of all naturally produced carbon dioxide emissions come from ocean-atmosphere exchange. Plant and animal respiration, and decomposition are also major contributors. Livestock farming and agriculture contribute greenhouse gases from fertilizers to cow flatulence to animal waste. In a bit of a convoluted process, U.S. corn is produced with the help of fertilizers and fossil fuels burned by farm machinery and converted to ethanol which is added to the country's vehicular fuel inventory.

What Has Changed and Who Should Do What?

Studies suggest that, since the Industrial Revolution, increases in the amount of greenhouse gases produced by humans have upset the delicate balance of what was a generally stable "carbon cycle". The burning of fossil fuels for heating, transportation and other purposes is generally considered to be the cause of this imbalance. So, significantly reducing or eliminating fossils fuel consumption will reduce the threat of global warming,

A Primer on Energy and Power Generation and Usage

Before we look for opportunities to reduce greenhouse gas emissions, let's first get a brief refresher on "energy" and "power".

In the simplest of terms, fossil fuels are burned to create energy which in turn is used to generate power and bring to life all those wonderful appliances and gadgets whose non-stop operation we have all quickly learned to expect 365/24/7. Understanding energy and power and how they are related helps us define opportunities to reduce power requirements and energy needs, and reduce or replace the energy generated from fossil fuels or other sources.

Energy

Energy is the capacity to do work; it's what must be transferred to an object in order for work to get done. Energy is expressed in terms of Joules. One Joule is a very small amount of energy. So, the kilowatt hour (kWh) is more commonly used.

- 1 kWh = 3.6 million joules

For example, burning one gallon of gasoline or diesel or fuel oil produces about 35 kWh of energy.

Power

Power is the rate of doing work. A power rating provides a measure of how much energy is consumed by a given piece of equipment.

An electric clothes dryer consumes about 3 kW of power

A 20 cubic foot fridge about 1.4 kW

By bringing time into the equation, we can assess how much energy is consumed for each packet of power generated .

Energy (kWh) = Power (kW) X Time (hrs)

So, burning one gallon of gasoline, which produces about 35 kWh of energy, will provide sufficient power to run an electric heater (that consumes 1kW of power) for about 35 hours.

This simple relationship shows us the connection between any source of energy (a gallon of gasoline) and almost anything that uses the power which that energy delivers. This could be your printer, the lightbulbs in a chandelier, the second refrigerator you have in your basement. When you buy a car that gets more miles per gallon, when you replace incandescent light bulbs with LED (light emitting diode) bulbs, when you get a new air conditioning unit with a higher SEER (Seasonal Energy Efficiency Ratio) you are tilting the Energy-Power relationship in a positive direction.

Less power needed, less energy used, fewer fossil fuels consumed to create the energy.

• *Of course, energy does not have to be created from fossil fuels. Alternative sources such as wind, solar and nuclear can be used instead of fossil fuels to provide that energy and power generation.*

The Future of Fossil Fuels (the Simple Story)

Fracking, and greatly improved secondary and tertiary oil recovery techniques (the latter known as Enhanced Oil Recovery), have markedly increased our ability to access oil reserves formerly considered out of reach. Despite recent reports that the world is quickly reaching the end of harvestable oil and natural gas, it now appears that significant inventories will persist for decades. Rarely are there discussions of "Peak Oil", much in vogue decades ago.

• *Peak Oil is the hypothetical point in the future where global oil production will reach its maximum and begin to decline.*

A secondary, related and equally important aspect of the recent growth of fossil fuel inventories is the likelihood that these full storage tanks will likely lead to continued low costs for both producers and consumers. This has important ramifications for efforts designed to reduce fossil fuel consumption.

Calculating Your Carbon Footprint

The term "Carbon Footprint" is used to characterize the extent of any one individual's use of fossil fuels and the corresponding generation of carbon dioxide gases. The process can be used to understand which activities and consumption factors are the biggest contributors to one's carbon footprint. Is it driving your car, is it the commercial jets you fly on for business and vacations, is it keeping your house at a comfortable 68 degrees all day every day? Then prioritize, from easiest to hardest, the changes in personal behavior and/or fossil fuel consumption that could reduce one's carbon footprint. Determine what is required to meet the extremely challenging, and in some cases downright ridiculous, per capita minimum carbon footprint guidelines being promulgated by international climate change committees.

The calculation of one's carbon footprint is a detailed, tedious task requiring a level of specificity few would care to undertake. The carbon footprint is calculated by integrating all of one's personal carbon contributions from the myriad of sources where they were expended. In addition to the seemingly straightforward contribution of gasoline to one's carbon footprint, one must also consider indirect emissions from, for instance, the refining of gasoline itself to losses in voltage as electricity moves through transmission lines to the more obscure categories of food, products and services. Yes, each food product (red meat, cereals, fruits, snacks) has its own unique set of processes that contribute indirect greenhouse gases emissions to the total. Waste itself. Examples include indirect greenhouse gas emissions from the trucks transporting the waste to the landfill, and the waste itself which creates methane, a highly detrimental greenhouse gas. Ultimately, calculating one's carbon footprint requires a numbing amount of information that only the most compulsive global warming zealots would be willing to collect and evaluate.

Climate Change – What's on Your "To-Do" List?

Now we are ready for you to help! To see what you can do to slow global warming. What you want to do, what you can afford to do, what the "system" will allow you to do. Let's see what's on your "to do" list. Individuals, the government and the marketplace all play a role in reducing greenhouse gas emissions. Each has incentives and methodologies. Each has supporters and detractors. Each has inhibitors that can slow progress. Each may need another's support to be successful. It's most of us intuitively understand that certain actions we take (or can take) contribute more (or less) to the creation of greenhouse gasses. Some are simple; some are more complicated; many involve nothing more than using energy more wisely.

Category	Easy/Low Cost	In the Middle	More Difficult and/or Expensive
Home Lighting	Turn off lights you are not using.	Use programmable light timers. Replace incandescent with LED light bulbs.	Install sensors that turn lights on only when they are needed.
Home Heating	(Permanently) turn the thermostat up in the summer and down in the winter. Turn off the a/c and open the windows on cool nights.	Use ceiling fans and/ or attic fans in conjunction with, or as an alternative to, air conditioning.	Add attic insulation. Install programmable thermostats.
Appliances	Hang clothes out to dry on a clothesline (just like Mom did in the 1950s). Power down gadgets you are not using. Take quick 5-minute showers.	Buy more energy-efficient appliances. Install on-demand appliance-specific hot water systems.	Add roof solar panels to produce hot water.
Transportation	Plot out the shortest distance for running errands. Wait until your car's gas tank is less than 1/5 full before refilling.	Let your high school student wait outside at the bus stop instead of sitting in the car with you with the engine running.	Seek out opportunities to rideshare.
Getting to Work	Carpool with a friend. Use mass transit.	Bicycle to and from work one or more days a week.	Telecommute one or more days a week. Buy a hybrid or electric car.
New Home Construction	Hot water only from on-demand units.	Add triple-glazed windows and attic and wall insulation.	Use of solar and wind to generate energy for running the house.

<u>Working Together – The Government, the Marketplace, and You</u>

Federal, state and local governments can help reduce the use of fossil fuels by requiring higher miles per gallon standards for automobiles and improved energy standards for appliances, and by developing regulations that require building standards that improve energy efficiency for new buildings. Since 2016, all federal agencies have been required to factor climate change considerations into their decision-making processes for granting permits. Local building codes could require enhanced features that improve insulation and reduce energy needs (triple glazed windows, double doors, and more attic insulation).

The energy utilized in hotels from lights and air conditioners left running in rooms which are unoccupied is extraordinary. Most of the energy completely wasted. Some hotels have keypads into which the room key must be inserted before electrical service can be restored. Even in the most unlikely of places, Kotzebue, Alaska, the Nullagvik Hotel, run by the local Native American community, has <u>state-of-the-art motion sensor technology</u> to turn lights on and off as needed.

<u>The Inhibitors – They're Out There and They're Big</u>

For all the initiatives to reduce fossil fuel consumption and greenhouse gas emission there are significant hurdles that have to be overcome to make any sizable dent in CO_2 emissions. Certainly not anywhere near the reductions being proposed by international global warming committees. These inhibitors include the low cost of energy and the high initial capital cost of alternative energy sources, unwillingness on the part of people to give up personal time or activities, opposition to government subsidies, personal choice clashing with government regulations, and indifference to, and general difficulty in understanding, how global warming affects "me". And why do I need to worry about it right now. Here are some examples that come to mind. There are many others.

Greenhouse Gas Reduction Solution	Inhibitor
Turn off lights and gadgets when not in use	Low cost of energy
Replace incandescent light bulbs with LED	High initial cost of LED bulbs Advertised 20-yr life not proven
Electric cars, solar panels, windmills	High capital and installment costs Limited travel range for electric cars Uncertainty about useful life, reliability and maintenance requirements in the absence of any credible long-term experience
Carpool to work	Limits scope of each carpooler's personal activities such as playing tennis before work and/or attending children's after school games
Take mass transit to work	Dependent on bus/train timeliness Walk/ wait in all kinds of weather May extend time required to get to work

Regulations that require shopkeepers to keep all exterior doors closed	Opposition to federal regulations Shopkeepers contend open doors are inviting, encourage business and improve profitability. Why should the government care if the shopkeeper pays his/her energy bills
Voluntary program that encourages personal commitment and independent actions to reduce energy consumption	Suspicion that others are not making the same sacrifice Uncertainty that actions are really making a difference

Each has incentives and methodologies. Each has supporters and detractors. Each has inhibitors that can slow progress. Each may need another's support to be successful. It's complicated.

CHAPTER 7 — WEATHER FORECASTING

Weather forecasts are prepared to provide the public with information about upcoming weather, whether it's for this afternoon or the upcoming winter snow season. Simply stated, the weather forecast comprises the expected changes that will take place over the coming timeframe in a location's temperature, sky condition, precipitation, and wind speed and direction. Depending on the season and where you live, you might also get wind chill, dew point and heat index.

So, by definition, this process of forecasting weather is not a simple one. In fact, it is incredibly complex.

NOAA's National Weather Service (NWS) offers a smorgasbord of weather forecasts for a variety of audiences and reasons. In the broadest sense, one can divide these forecasts into Long Range and Short Range forecasts. If we consider short term Special Weather Statements prepared for instance to warn the public about rapidly occurring, unexpected black ice, snow squalls or dust storms, the lower end of short term forecasts can be measured in minutes. At the other end of the forecast timeline would be the multi-month Seasonal Forecasts.

Each forecast has a purpose and a set of factors that define that purpose. Ultimately most short term weather forecasts are prepared to provide the general public with information on threats to their lives or property, and what to do to decrease such risks, such as taking cover when a tornado touch down has occurred nearby. Such a forecast would almost without question be accompanied by a severe weather warning. If more time is expected before tornadoes arrive, a weather watch may be sufficient. The urgency associated with any potential or possible threat defines the timing associated with such a risk. Typically, the shorter the timeframe the more imminent the risk, and vice versa.

On the other end of the spectrum, long range weather forecasts provide either scenarios about impending weather risks more than a week out, or for the longer times yet, information with which to plan activities such as how much sand and salt a municipality should stockpile for the upcoming winter.

The greater the time horizon the lower the forecast accuracy.

The challenge is to develop forecasts sooner and more accurately.

Table 1 below shows the time frames associated with the various forecast names used by the National Weather Service.

Name		Time Frame
Short Range Weather Forecast		12-48 hours
Mediums Range Weather		3-7 days
Long Range Weather Forecast		8-14 days
Seasonal Outlook		Each Quarter of the Year
Hurricane Outlook		Every April or May

First let's take a look at each of the timeframes in some more detail, starting first with the longer range weather forecasts and proceeding to the shorter range forecasts.

Hurricane Forecasting

The hurricane season in the Atlantic Basin runs each year from June 1 to November 30, although there can be, and many years is tropical storm development before and after these dates. NOAA's seasonal hurricane forecast provides probabilities for the development of tropical storms, but does not predict how many storms will hit land or where they will strike. Without the landfall probabilities and locations, this exercise becomes merely a question of how many, not the valuable where and how bad.

But the general public seems to like the estimates which do profile the relative changes in tropical storm strength from year-to-year. A term used primarily in academia or meteorological settings is "Accumulated Cyclone

Energy" (or ACE) which is the sum of the energies of all the Atlantic Basin's individual storm each year. This metric does provide a quantitative indicator of combined annual tropical storm strength, by hurricane season.

In addition to NOAA, several other influential entities make estimates of the number of named tropical storms, hurricanes and major (Category 3+) storms that will occur this year. These include The Weather Channel (TWS on the table below) and AccuWeather (ACCU). Perhaps the most unusual contributor is Colorado State University in Boulder CO. Yes, a landlocked northern school studying hurricanes.

<u>TABLE 2</u>

	2022	NOAA	CO	TWC	ACCU
Named Storms (>39 mph)	9	13	13	15	11-15
Hurricanes (>74 mph)	5	6	6	7	3-6
Major Hurricanes (>111 mph)	1	2	2	3	1-3

There is excellent historical hurricane history, in the form of data, that goes back to 1850. How many hurricanes? How big? Category? When? Where? Landfalls? And so forth. All in the public domain. For example, a map tells us that, since 1900, 73 Cat 3-4-5 hurricanes have made landfall on U.S soil and which states were affected.

And a wonderful interactive tool has been developed that defines the probability of landfall and attainment of specific wind velocities for each of the 205 "shore" counties in the 18 East and Gulf Coast states. Data is provided for named storms, for Cat 1 and 2 hurricanes, and for intense hurricanes (Categories 3, 4 and 5) with winds greater than 115 mph.

With the amazing history that has been collected you can even go back further than 1850 to read about major East Coast storms before they were called hurricanes. You can add and subtract the number of hurricanes of a certain size by year or whatever you like. Find a "hurricane" that took place way before hurricanes were "in", like the hurricanes of 1703, 1716 or September 27, 1727, that affected eastern New England with "A Great Rain and Horrible Wind". Or wonder, or research, why the storm of 1850 is called "The Triple Storms of 1850". It's an amazing set of data.

<u>Tracking Hurricanes</u>

Shortly after a possible tropical system begins to take shape forecasters begin to look at factors that will influence the speed and direction of the potential storm that may help them define the current and future positions of the jet stream, and high and low pressure systems, that can play dramatic roles in the storm's intensity, and its direction and speed. As storms begin to approach populated areas, hurricane experts will rely increasingly on modeling. And since obtaining sufficient and accurate meteorological information over the open waters of the Atlantic is problematic, hurricane hunters will begin to play an indispensable role, providing data and information literally unobtainable any other way.

Ultimately a small number of storms will approach landfall. And where exactly that landfall occurs can have dramatic consequences for those in its path. Despite the uncertainty of so many features of hurricanes, there is one certainty, and that is that the exact location of landfall clearly defines who will experience perhaps significant or catastrophic damage, and who will have markedly fewer consequences.

Let's assume a hurricane is making landfall at a spot on the map. Let's say this landfall is on the East Coast, and assume the storm is coming ashore with a westerly or northwesterly trajectory. To get this point across, let's assume the exact landfall of the eye of the storm is 65 miles southwest of Wilmington, NC, near Myrtle Beach, SC.

Almost all of the damage from this landfalling hurricane will occur in the storm's so-called <u>"northeast quadrant"</u>; that is to the north and east of the spot where the eye comes ashore. It is here that potential damage is at its worst because first, the winds will be at their highest, flowing ashore with no impediments. And second, more importantly, this is the area in relation to landfall where a potentially powerful and destructive storm surge, a wall of water several-to-many feet high, propelled by those same winds, comes ashore. Should this be a major Category 3+ hurricane, the surge will literally erase everything in its path, people and property. Wilmington could suffer significant damage in this scenario were it to be a major hurricane.

Once the hurricane comes ashore, the circulation of the hurricane is immediately slowed by the natural (hills, rocky shore) and manmade (buildings) features of the landscape. So, the winds that come all the way around the storm and reach the area to the southwest of the eye have greatly reduced speeds. Here there is no storm surge.

Were our hypothetical hurricane to come ashore instead 75 miles up the coast northeast of Wilmington, say in Morehead City, NC, then Wilmington would be in the just-described quiet "northwest quadrant" and experience a significantly reduced impact. Where you are in relation to the exact location of landfall is extremely important. In the scenario just described, a landfall difference of just 140 miles either side of Wilmington, NC will lead to a markedly different result for the city.

NOAA has accumulated an impressive array of data from hundreds of landfalling hurricanes that defines the so-called "Landfall Position Forecast Error" for various forecast lead times. In just the last decade, forecasters have gotten considerably better at pinpointing the ultimate spot where landfall occurs.

Seasonal Outlook

Seasonal forecasts are forecasts of average seasonal conditions (temperature and precipitation) over a specific region for usually three months in advance. This is possible because of the slowly changing nature of climate systems. For example, assume water temperatures are currently quite a bit above historic average in the Gulf of Mexico. Because ocean water temperatures change very slowly (particularly when compared to the Earth), there will not be a dramatic changes in the seasonal outlook three months from now.

Seasonal forecasts primarily use climate models to predict what the seasonal rainfall might be 90 days away. Methods include Dynamical Forecasting, which uses 3-D climate models that alert the forecaster of possible long range jet stream or air mass movements. Such high level changes could signal more substantial changes to temperature and precipitation down the road.

The official 90-day outlooks are issued once each month near mid-month at 8:30 am Eastern Time.

TABLE 3

What a Seasonal Forecast Is	*What a Seasonal Forecast Is Not*
Information about average seasonal conditions	Information about conditions on day-to-day variations or extreme weather
Forecast over a large region	Forecasts with small scale spatial detail
Shifts in probabilities	Definite

Long Range Weather Forecasts (8-14 days)

NOAA's Long Range forecast has a timespan of 8-14 days. Interestingly enough the 14-day forecast period is four days longer than the generally accepted NWS position that 10 days represents the longest lookahead time for which we can reasonably expect reliably accurate forecasts. A study published in the April 2023 Journal of the Atmospheric Sciences said, "A skillful forecast lead time of midlatitude instantaneous weather is around 10 days, which serves as the practical predictability limit." Because of the relatively short look-ahead time, these forecasts tend to be generally accurate, though more so for temperature than for precipitation.

On the shorter end of the Long Range Forecast timeline, eight days is at the point where forecaster can begin to see large (and perhaps familiar looking) features on "the map". Trends involving the jet stream, air masses, pressure systems and fronts are emerging and possibly of value to forecast us. The forecasts are produced by the Climate Prediction Center (CPC), a branch of the National Weather Service.

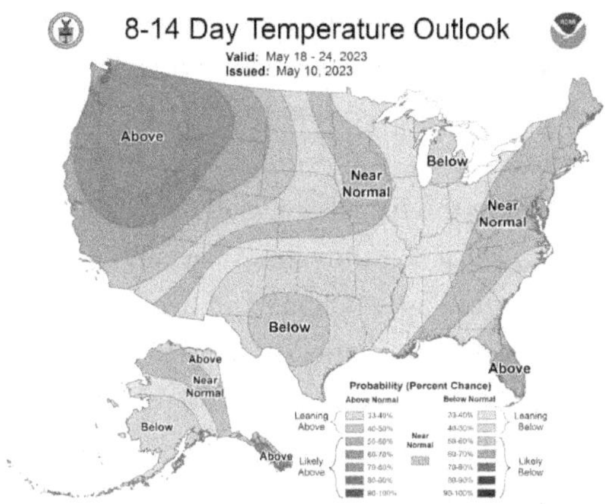

<u>Table 4 Medium Range Weather Forecast</u> (3-7 days)

Surface maps are being drawn with increased confidence. Models begin to recognize areas of low pressure. It's close enough now that forecasters can see the potential for an area of energy in the Gulf of Alaska to become a nor-'easter and scoot up the Eastern Seaboard. The "Five Day Forecast" is often the last weather information provided on TV.

At this point in the forecasting process, the forecasters have an excellent idea of the issues they will face, and with a high degree of confidence. Surface maps show locations of high and low pressure systems, storm formation, temperatures and probability of precipitation are much clearer. It is in this realm of forecasting (3-5 days) that the greatest improvements have been made over the last two or more decades

<u>Short Range Weather Forecast (12-48 hours)</u>
Short range weather forecasts have by definition a horizon of 18-48 hours. <u>The Short Range is by far the most accurate, informative, interesting, detailed and sought-after weather forecast.</u> The most common forecast within this short range category is the so-called "36 Hour Forecast". which provides the forecast for today, tonight and tomorrow. Meteorologists who are assigned the task of developing and issuing weather reports depend heavily on data, modeling, radar, plus bit of intuition and some experience from colleagues

Meteorologists responsible for the short term forecast need to be ready to respond to unexpected weather events occurring in real time that present motorists and pedestrians with risks not in the forecast. Examples include fog, black ice and snow squalls.

No second thoughts at this time. Forecasters can reliably provide high and low temperature within a degree Fahrenheit, provide accurate likely wind, and rain and snow totals, as well as wind chill and heat indices. Day of the overnight and early morning forecasts are detailed and generally accurate. Forecasters have worked what-ifs and other long range weather alerts and forecast.

The <u>inclusion of radar in the meteorologist's toolkit greatly increased the confidence in, and accuracy of, short term weather forecasting</u>.

There is hardly an organization anywhere that doesn't need to know the information contained in a weather forecast. So, who needs to know about the weather? Airports surely do but then so do farmers, hospitals, wedding planners, water resource managers, ski resorts and school systems, just to name a few.

The Early Days – A Growing Urgency to Improve

It was becoming increasingly clear that more data, better tools, stronger leaders were needed before weather forecasting could be viewed as anything more than educated guesswork. Three storms decades apart caught almost everyone's attention and cries for improved weather forecasting got stronger and louder

The Great Blizzard of 1888, one of the most significant snowstorms in recorded history to affect heavily populated areas, with snow depths in New York City and nearby Connecticut of greater than 2-4 feet, and drifts covering second story windows. The weather forecast that day was for <u>fair skies and warmer temperatures.</u>

A little more than a decade later, on September 8, 1900, a major hurricane rolled into Galveston, Texas from the Gulf of Mexico. Perhaps as many as 12,000 people were killed—hardly anyone was left to make the count. <u>And no one knew it was coming</u>! The cries to get their act together, to develop a knowledgeable Weather Bureau that could provide warning and save lives.

And then, the Great New England Hurricane of 1938 struck the southern coasts of first Long Island and then Connecticut, causing significant damage and loss of life. At that time, the Weather Bureau was operating an unwieldly organization that required responsibility for any hurricane moving up the East Cost to be transferred from the Bureau's Jacksonville, Florida office to the New York City office on the run. The 1938 hurricane was a fastmoving storm that fooled the organization. The younger meteorologists were predicting much less, if any, turn out to sea, with a possible landfall somewhere between New Jersey and Cape Cod. The elders prevailed; the young weathermen were told to go home. Instead, a fair day with light winds was forecast.

It was becoming clear to scientists, engineers and many others that weather forecasts would never be able to effectively serve the public unless two gaps, which had continued to elude meteorologists, could be filled. Those two gaps were:

1. How to analyze the ever expanding data, information and observation of meteorological features
2. How to "see" the weather

The first to propose the fundamental ideas behind numerical weather forecasting was Norwegian physicist and meteorologist Vilhelm Bjerknes, who foresaw the possibilities as early as 1904. In 1922 English mathematician and meteorologist Lewis Richardson converted Bjerknes' concepts to equations, only to find out that it was impossible to do all the calculations in anywhere near a reasonable time, estimating it would take 26,000 people working 24/7 to make daily numerically based weather calculations and predictions.

The roadblock was gradually removed when the forerunner of modern computing was devised in the early 1940s.

With the ability to now do calculations beyond (at the time) anyone's wildest dreams, it was left to Princeton meteorologist Jule Charney to "write the numerical code" and let the computers do their thing. The first successful weather forecast using numerical prediction techniques was crunched by the "super computer" (of its time) at Maryland's Aberdeen Proving Ground. By 1955 the Weather Bureau was issuing daily forecasts at least in part supported by numerical techniques.

Data Supports Weather Forecasting

Every weather forecast starts by using people, equipment and instruments to collect data. Weather data is collected in all sorts of ways and for all sorts of reasons. The most complete and accurate data is collected at the National Weather Service's 122 Local Weather Forecasting Offices (LWFO's) which measure, distribute and store data in data banks supplementing the LWFO's are 500 or so Automated Surface Observing Stations (ASOS's) sprinkled around the country. These unmanned stations measure temperature, barometric pressure dew point, wind speed and direction. "Accumulated Energy" precipitation, sky conditions and visibility, temperature (high, low, high minimum) send all this data back to the LWFO's.

To obtain data on upper-air wind speed and direction, the balloons can be tracked by radar or satellite-based GPS, sometimes for up to two hours.

With this immense database, weather maps can be drawn, and revised periodically, to show areas of similar temperature, wind speed and direction, equal barometric pressure, and so forth. And using this data and information, combined with personal experience and perhaps a little bit of intuition, forecasters can quite confidently prepare a 24 or 36 hour forecast with a generally high likelihood of accuracy.

The National Weather Service also maintains a system of marine weather observation sites including 90 buoys and 60 land-based coastal observing stations. There are 92 locations in the U.S. that launch nigh altitude balloons fitted with high altitude balloons. Ascending at 100 feet, they send back atmospheric pressure, wind and humidity data.

Using Data - Models

Models consume and create prodigious amounts of data to develop numerically based weather forecasts, run the models, and review and continuously improve the process. Ships at sea also are invaluable in proving data almost impossible or hard to obtain in any other way, such as wave height and ocean water temperature. Pilots can also help fill in missing data need for dodging severe wind events like downdrafts and clear air turbulence.

But the award for best technique to help fill in missing data from, for instance the Commonwealth of Virginia, is the nationwide army of 11,000

Cooperative Weather Observers who report daily or upon demand their locale's temperature, rain and snow, sky conditions and so forth. This dedicated group of volunteers broadens the database, filling in the holes of official data gathering around the U.S. From a long term perspective, these observers play a major role in defining and maximizing the accuracy of the country's climatological history, making sure the records are as accurate as possible.

Major International Models

Weather forecasting models are typically run on huge government computers, or by institutions such as the Organization of European States. There are a small number of computer models used around the world for short and medium range forecasting.

- The American Global Forecast System GFS (run by the National Weather Service)
- The European Model ECMWF (European Center for Medium-Range Weather Forecasting)
- The U.K. Model UKMET (United Kingdom Meteorological Centre)
- The Canadian Model CMC (Canadian Meteorological Center)
- The Japanese Model JMA (Japanese Meteorological Agency)

All are "global", "dynamical" models, meaning they can solve mathematical equations governing the behavior of the atmosphere at every point on the globe. Because models are expensive and time-consuming to run, operators may have to choose between a high resolution, short term forecast and a lower resolution, long term forecast. It takes about 2 hours to run a global model on the world's most advanced supercomputers. The global models prepare forecasts with different time horizons, ranging from six days for the UK model to 16 days for the U.S. Global Forecast System, although with significantly decreased accuracy after 10 days.

And Radar to "See" the Weather

Radar is an acronym for Radio Detection and Ranging. It uses radio waves to detect objects in the air. The science was discovered in the late 1800s. At first it wasn't clear how to use this new technology. World War II changed that. After the war, an increased understanding of radar's capabilities made it natural for watching the weather.

When short pulses of electromagnetic waves intercept precipitation, part of the energy is reflected back to the radar. The strength of the pulse returned to the radar depends on the size of the particles, how many there are, and what state they are in. Computers then create integrated weather radar images. These are the radar maps so common today on the internet and TV. Depending on the intensity of the precipitation, different colors will appear on the map.

- Rain from lower to higher intensity characterized by shades of green, yellow, orange and red
- Mixed precipitations (sleet, snow, rain) by pink and red
- Snow by light and dark blue

By 1955 the Weather Bureau was issuing daily forecasts at least in part supported by numerical techniques.

The Global Forecast System, the computer model run by the National Weather Service, only produces forecasts as far as 16 days in advance. The consensus is that eight days represents the time period within which NOAA has the most confidence, The highly regarded "European Model" only forecasts out as far as 10 days. These two systems are perhaps the most sophisticated weather modeling programs on the planet. And even they admit unpredictability in forecasts more than a week's time in advance.

Newspapers – The First Messengers

The British Meteorological Office began issuing daily weather forecasts in 1860. These were shortly thereafter published in the *London Times*, and so it began.

By 1875 the *London Times* was publishing weather maps showing prevailing winds, barometric pressure and temperatures throughout the British Isles and portions of Western Europe. Other countries soon followed suit and by the last decade of the nineteenth century weather forecasts, often accompanied by maps, were a fixture in many daily newspapers.

Of course those forecasts were often of questionable accuracy. Nonetheless, the process was up-and-running—messengers were distributing weather forecasts to an interested populace. And although printed newspapers command a smaller and smaller circulation, weather maps can still be found in mostly newspapers.

A nowhere else can you get the scope, detail and perspective provided by the "Weather Section" of a major newspaper.

- As an example, every day the *Washington Post's* weather section you can find

1. *A quick synopsis of the D.C. Metro's 24-hour forecast*

2. *The 6-day forecast for Washington including graphical representation and the following info—sky condition, high and low temperature, "feels like" temperature, probability of precipitation, wind speed and direction*

3. *A regional map extending from Harrisburg, Pennsylvania to Kitty Hawk, North Carolina with forecast sky condition and temperatures*

4. *Forecasts for pollen, UV and air quality*

5. *Three-day forecasts for the Blue Ridge, the Atlantic Beaches (ocean temps provided) and local waterways like the Chesapeake Bay*

6. *Tide tables, phases of the moon, and rise and set times for the planets*

7. *A bar graph depicting the average, actual and forecast range of daily temperatures and precipitation for the past 5 days, today and the next 9 days for all 3 airports (IAD, DCA, BWI)*

8. *A weather map of the United States, southern Canada and northern Mexico containing the location of high and low pressure systems, fronts, and precipitation fields; colored high temperature zones; and high and low temperatures for about 30 major cities, and finally*

9. *Today's and tomorrow's forecast high and low temperatures and weather conditions for about 150 U.S. and international cities*

Admittedly, some of the information is more closely aligned with astronomy, human physiology, oceanography and geography than it is with meteorology. But one must admit it's an impressive collection of information. Other newspapers in other geographic locations might include ski reports, reservoir and lake levels, and degree days.

What else could you need? Well, as it turns out—one very important feature—*timeliness!* If you want to know what's going on with the weather right now, it's the internet or the trusty local AM or PM radio station you will rely on. The newspaper is frozen in time.

Radios Play a Major Role in Weather Transmission

The British Meteorological Office is the agency credited with the first radio transmission of a weather forecast, a marine forecast in 1911. In the United States, public radio weather forecasting began in 1925 at Boston's WEEL. Radio remains a powerful conveyer of weather information, allowing virtually instant reporting of current conditions and updated forecasts. NOAA Radio also plays a major role in conveying weather information to the general public.

This is a particularly important service in rural/remote towns/homesteads far from the more accessible information available in more populated areas. *NOAA Weather Radio All Hazards* (NWR) is a nationwide network of radio stations broadcasting continuous weather information directly from the nearest NWS office. Official forecasts, advisories, watches warnings are available at all times. NWR is comprised of over 1000 transmitters, covering all 50 States and adjacent coastal waters. All but about 5% of the 300 or so million inhabitants of the U.S. can receive NOAA's signal.

Access to Weather Data and Forecasts Now Almost Instant

Washington D.C.'s WTOP radio, and many others like it, provide weather information and current forecasts every 10 minutes. Despite the skepticism that accompanied the launch of 24- hour "nothing but weather" cable TV stations, The Weather Station, AccuWeather and others have proven to be successful, with millions of dedicated viewers. For most of us, getting a read on tomorrow's weather, either the night before or the morning off, provides input on what type of clothes to wear, whether or not to bring an umbrella, and so forth. Nowadays almost everyone has at least several weather apps for their iPhones, providing access to current data and information only dreamed about years ago (even by professional meteorologists). Not only can you get the temperature in your backyard a minute ago, but you can also get it at your son's house halfway across the country.

Apps let you track an oncoming thunderstorm, count the number of lightning strikes in a nearby county, and assess whether blizzard conditions can be expected in the next 48 hours. A special radio receiver is required to pick up signals broadcast at seven specific frequencies. The radios, which can be programmed to "beep" when watches and/or warnings have been issued, are readily available and inexpensive. Numerous lives have been saved as a result of this messenger system. A number of experimental broadcasts were made by the BBC and America's Dumont Television Network in the 1930s and 1940s. But it wasn't until the late 1970s that the precursors of what we associate with TV weather "personalities" today, features like satellite images, detailed and graphically appealing weather maps, and radar depictions of precipitation, began to emerge. Today, most major TV stations in America have a team of meteorologists. During the days preceding a major storm, more people turn on the TV to watch the weather than for any other reason.

<u>Forecasting Challenges</u>

It's a really big job for the National Weather Service to meet its mission of protecting the general public from meteorological threats and hazards. Every day!

Not only do you need to develop and maintain processes and systems, have access to data and the models that use it, and meet deadlines, you need to have the people, smart people, to pull it off.

Despite having access to incredible data, despite having RADAR and satellite imagery to assist with storm movement and speed, despite having lived in the same area for many years, preparing a fresh 24–36-hour weather forecast remains a difficult though gratifying experience. The atmosphere is a turbulent, confusing space. Things don't always go as expected. Sometimes in the car you get a good long time to see adverse weather forming ahead or besides you; other times a hail-producing cell forms right above your car.

Forecasters must assess and consider many features and understand a lot about how jet streams force weather. But it's the non-linearity that forecasters can't really comprehend without the aid of sophisticated weather prediction model. The complicated nature of the relationship between jet streams and their impacts on the weather is generally understood. What forecasters can't really comprehend is the complex and complicated roles played by jet streams, low and high pressure systems, fronts, temperature, humidity, and barometric pressure.

Forecasters must consider navigate through a number of factors which will be included in their forecasts report (see NOAA report from Dulles; Airport to see the breadth of this data), most of it with a few exceptions not just high and low temperatures. Remarkably, this updating process finishes in the morning of the previous day. Exceptional.

Among the factors that make up a particularly effective forecast are cloud cover, type of cloud, sky condition and type of snow. Also included are estimating the condition of the sky (cloud coverage in eighths of a half-sphere), cloud type (clues for weather), precipitation intensity, and last but not least probability of precipitation.

<u>Getting the Right People – Forecaster Requirements</u>

Interestingly enough, many TV weather personalities do not have college degrees in meteorology. This apparent dichotomy works as long as the stations pass along, for instance, Severe Thunderstorm Warnings, Hurricane Watches and the like. These watches, warnings and advisories are issued by NOAA's Severe Storms Center in Kansas City, Missouri. This notification process is the sole responsibility of NOAA. Of course, private entities can make their own forecasts that may feature the Warning released in Kansas City. The they just can't distribute their forecasts to the general public.

So, do you want to work for the National Weather Service?

You will need a degree in Meteorology, Atmospheric Science or Hydrology with required proficiency in some combination of the following subjects.

- *Atmospheric Dynamics, Analysis and Prediction of Weather Systems, Physical Meteorology, Remote Sensing of the Atmosphere, Instrumentation, Physics, Differential Equations, Physical Hydrology, Physical Climatology, Aeronomy, Advanced Electricity and Magnetism, Statistics, Physical Oceanography, Heat Transfer, Advanced Thermodynamics*

<u>Describing Sky Conditions</u>

What's the difference between partly cloudy and partly sunny? Or what about a forecast that says cloudy, when another says mostly cloudy? Is there an appreciable difference, significant enough to have two descriptors? Are overcast and cloudy the same? How about clear, sunny and mostly sunny?

As it turns out, these terms are dependent on the portion of the sky covered by clouds expressed as a fraction as eighths, and on the time of observation—night or day.

Term	Means	Same As	But Different Because
Sunny	No clouds	Clear	Sunny used only during daylight hours
Mostly Sunny	When 1/8 to 2/8 of the sky is covered by clouds	Mostly Clear	Most Sunny used only during daylight hours
Partly Sunny	3/8 to 5/8 covered	Partly Cloudy	Partly Sunny used only during daylight hours
Mostly Cloudy	6/8 to 7/8 covered	Considerable Cloudiness	
Cloudy	7/8 or more covered	Overcast	Overcast used for aviation observations

Snowfall Characterizations

TABLE 6

Type of Snow Event	Description
Snow Flurry	Intermittent, short duration, trace accumulation at most
Snow Shower	On and off intervals, varying intensity
Light Snow	Steady; intensity low enough that visibilities stay at or above 1/2 mile
Snow Squall	Intense, limited duration, periods of moderate to heavy snowfall, significant accumulations possible
Heavy Snow	4" or more in 12 hours and/or 6" or more in 24 hours

<u>Probability of Precipitation (PoP)</u>

No single meteorological parameter is as misunderstood, or difficult to explain, as Probability of Precipitation (PoP). The National Weather Service definition is

"Probability of Precipitation (PoP) is the probability that precipitation will be reported at a certain location during a specified period of time"

- The forecaster considers all the precursors that could lead to rain during the upcoming forecast period.
    - o A low pressure system, convection, and/or orographic features
- The forecaster assesses the state of the atmosphere regarding the likelihood, duration, coverage and intensity of rain to be expected in the upcoming forecast period over the forecast area for which it applies
- The Probability of Precipitation has the following features and expectations
- *The likelihood that precipitation will occur, expressed as a percentage*
- *A measurable amount of precipitation of at least one hundredth of an inch (0.01")*
- *A defined forecast area for which the forecast is applicable, for instance Arlington County, VA*
- *A specified period of time for which the forecast applies*
    - ■ *Forecasts are updated frequently, but the horizon of applicability for the short term forecast still tends to be just a few hours ahead*
- So when the forecast reads: *"The Probability of Precipitation is 40%"* it means
    - o There is a 40% probability that precipitation of at least 0.01" will occur at some place in the forecast area during the specified forecast period
- As important as what PoP is, is what PoP is not. PoP does not mean
    - o That 40% of the forecast area will receive precipitation during the forecast period
    - o That precipitation will be falling on the forecast area 40% of the time during the forecast period
    - o Here are some common words used to describe the various statistical probabilities of precipitation

Table 7

POP	Term
0%	None
10%	Isolated
20%	Slight Chance

Most clouds do not yield rain
And, if there are no clouds, it can't rain

Cloud Formation and Classification

Clouds are really nothing more than trillions of teenie-weenie ice crystals and water droplets. When the concentration of these particle becomes sufficiently high the water vapor will become visible as a cloud.

The World Meteorological Organization (WMO) has endorsed only a dozen or so cloud types, a remarkably low small number considering the size of the Earth and the diversity of its climate.

Although clouds may be virtually omnipresent, few people can actually describe or identify the differences between cloud types. Clouds are an indispensable contributor to the water cycle, generate shade that protects humans and other mammals from the intense heat of the sun's rays and contribute to the beauty of sunrises and sunsets.

The National Weather Service (NWS) groups clouds according to the altitude level at which its base is located.

- High Level clouds (made almost entirely of ice crystals) and positioned more than 20,000 feet above the Earth's surface)
- Medium Level clouds (which form at heights between 6,500 and 20,000 feet)
- Low Level clouds (with bases below 6,500 feet)

Cloud Naming Conventions

In 1803, an Englishman, Luke Howard, developed a cloud classification system that still serves as the model for the international system for naming and defining cloud types. Each cloud type's name is composed of a one or more words or associated prefixes or suffixes that identify

Altitude - location of the cloud's base expressed in feet above the Earth's surface

- *Appearance* – is the cloud predominantly stratus, cumulus or nimbus
- *Propensity for rain* – is the cloud a rainmaker or not?
- Specific words are then used to distinguish clouds by their appearance and characteristics:
    - o Cirrus – wispy, feathery, and composed entirely of ice crystals
    - o Stratus – low level clouds characterized by horizontal layering
    - o Cumulus – puffy, cotton-like or fluffy features

- *Nimbus* is added to indicate that the cloud produces precipitation.
    - o The clouds associated with precipitation are (not surprisingly) dark, gray, featureless, and thick enough to block out light from the moon and the sun. These clouds usually produce continuous rain or snow, but not thunder and lightning.

<u>Cloud Types</u>
- Using the combination of appearance, altitude and ability to create precipitation, a variety of cloud types can be identified.
- Here are some cloud types and their respective altitudes recognized by the NWS and the World Meteorological Organization (WMO):
    - o Cirrus – High Level
    - o Cirrostratus – High Level
    - o Altostratus – Mid Level
    - o Altocumulus – Mid Level
    - o Cumulus – Low Level
    - o Cumulonimbus – Low Level
- So cloud names are in most cases "hybrids" made up of pieces from the several categories that define altitude and appearance.
- Here are a few examples:

- *Cirrocumulus* _ Cirro means wispy and at high elevation; *cumulus* means puffy or fluffy features, and since *nimbus* is not part of the name, we should not expect any rain.
- *Cumulonimbus* – *Cumulus* means puffy or fluffy characteristics *nimbus* means rain-producing; so this is a cloud with dark black rolling structure likely to produce rain or snow.
- An amazing feature of cloud naming is that <u>only five words</u> or phrases in total (*cirrus, cumulus, stratus, alto and nimbus*) <u>are needed to define cloud characteristics</u> including altitude, appearance and rain/rain likelihood. In essence cloud names are simply hybrids that combine the various data on altitude, appearance and rain into a single word like *Stratocumulus*.

Take a Closer Look at NOAA A

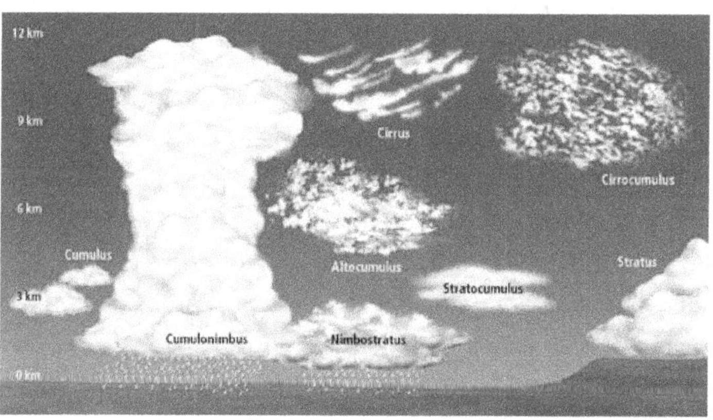

NOAA is the United States' "National Oceanographic and Atmospheric Administration". Organizationally the National Weather Service NWS report through NOAA to the U.S. under Secretary of Commerce. The NWS has over 4,000 people working in scientific, technical and administrative positions. Its primary mission is greater understanding and application of technology in climate, weather, the ocean and coasts.

Let's take a look at some of the work ongoing at NOAA in the area of hurricanes, discussed in the chapter "Storms" . To improve hurricane intensity forecasts, NOAA needs more data. In 2021, NOAA formed the first integrated ocean observations team to coordinate the use of different observing platforms including gliders, drifters, floats, small uncrewed aircraft systems UAS, and more. The group leveraged existing observations and so operated in extreme conditions, battled 50-foot waves and winds of more than 120 miles per hour to collect critical scientific data.

The Service Assessment – NOAA's Internal Audit

NOAA periodically conducts so-called Service Assessments to evaluate its performance in recognizing, forecasting and issuing appropriate advisories, watches and warnings for major meteorological events. Special attention is given to those that have the potential to, or already did, affect human safety, infrastructure/or residential and commercial buildings.

NOAA's mission is understanding and predicting changes in climate, weather, the ocean and coasts.

Science at NOAA is the systematic study of the structure and behavior of the ocean, atmosphere, and related ecosystems; integration of research and analysis; observations and monitoring; and environmental modeling. NOAA science includes discoveries and ever new understanding of the ocean and atmosphere, and the application of this understanding to such issues as the causes and consequences of climate change, the physical dynamics of high-impact weather events, the dynamics of complex ecosystems and biodiversity and the ability to model and predict the future states of these systems. Science provides the foundation and future promise of the service and stewardship elements of NOAA's misses.

Such an assessment was conducted following the historic Derecho of June 29, 2012. A Derecho is, in layman's terms, a very fast-moving storm front pushing an energized air mass in front of it at speeds of 60 mph or greater. Derechos occur infrequently in the Mid-Atlantic. But the 2012 event was a dangerous windstorm.

The conclusion of the Service Assessment was that the various agencies and tools responsible for identifying and forecasting such potentially dangerous weather events were slow to do so. *"The forecast model provided little assistance in forecasting this event more than 24 hours ahead of time."*

Remarkable Successes

On average, a five-day weather forecast of today is as reliable as the two-day weather forecast was 20 years ago .

But amidst the occasional misses, there are indeed many remarkable successes. How about Hurricane Sandy? Bam! A spot-on forecast for a left turn. Such a track was virtually unprecedented, with "typical" paths running parallel to the Shore and/or veering to the northeast and out to sea, making such a forecast not only unorthodox but difficult to comprehend. But and despite all that, and the fact that inherently the hurricane/tropical storm is the weather feature whose behavior is most difficult to predict, the forecast was astoundingly accurate.

Or how about the Blizzard of 1996? Forecasters correctly predicted that a wave of energy off the coast of Oregon would slide across the country, get organized in the Gulf of Mexico, move up the eastern seaboard, pick up energy and evolve into a "Superstorm" that would drop over 30" on snow on New Jersey? Yes, they got it exactly right.

The Bottom Line – Pretty Darn Good

Day in and day out, NOAA fulfills its mission of "informing the public of upcoming weather in order to protect people and property, and to enhance the overall economic performance of the nation".

- Every year the National Weather Service issues more than 734,000 weather, aviation, marine and fire forecasts—2,000 a day.
- These forecasts are comprised of standard 36-hour short range look-ahead and longer term 7-day prognostications.
- This costs each of us less than $10 a year!

These "standard" forecasts are complemented by Advisories, Watches and Warnings. 850,000 of these annually, most issued in distinct geographic locations, for example. *Watches and Warnings for Severe Thunderstorms and Flash Floods/ Gale Warnings, Small Craft Advisories Heat, High Surf and Fire Weather Watches.*

Imagine having the responsibility to predict the sky condition, humidity, high and low temperature, and wind speed and direction for every dot on the surface of the Earth. Because many of those dots represent a spot where somebody lives and/or is travelling to, it is a daunting task, but one that NOAA does quite well.

CHAPTER 8 — RISK AND WEATHER

This article will describe a process used to identify hazards, and assess, compare and reduce risks, a so-called risk assessment and management process. The premise is that such a process is well-suited for assessing and addressing risks associated with the climates in which we live; everything from a light snowfall to a Derecho. The process relies on a risk matrix to visualize and compare assessed risks. Risk management examines the assessed risks and seeks a set of cost-effective changes that reduce risk probability and/or consequences.

The study of risk is a complex, complicated, detailed and non-exact undertaking. We will be looking at just a very thin slice of risk "technology", and of course concentrating on weather-related risks.

What is Risk?

Risk is the probability that a situation involving a particular hazard will occur, coupled with the consequences if it does. A hazard is anything potentially costly, harmful or undesirable. Hazards lead to risks. The connection between a hazard and a risk is an event, usually referred to as a scenario. Scenarios address the "what if's?" of risk—what does or could happen, how frequently does it occur based on actual data and/or the best estimates and what can be done to decrease the risk associated with each scenario. The same hazard can lead to a number of different risks. For instance, water in a plastic cup on a table represents a low risk. But that same water spilled onto a slippery floor or as a component of black ice or a hurricane's storm surge represents an entirely different risk scenario. So, scenarios tie together the ingredients of risk (hazards, events, probabilities and consequences). There are many types of risk, including personnel safety risks, business risks, process safety risks, public health risks, security risks and environmental risks. Risk is everywhere. We encounter risks driving our cars, walking across the street, drinking alcohol, sky diving, being obese, cleaning leaves from gutters while sitting on the roofs of our homes, and so forth. And there are also weather risks.

Weather Risks

In the world of meteorology, the primary atmospheric hazards that create risks to people, their homes and businesses, infrastructure, and flora and fauna, are temperature, wind and water. Generally we live comfortably with these omnipresent atmospheric hazards. Take for example the following weather forecast: *"Partly cloudy with a 40% chance of thunderstorms and a high temperature of 80 degrees. Light northwesterly winds at 8-15 mph except higher gusts in thunderstorms between a tenth and a quarter of an inch with higher amounts possible in thundershowers."* Risks that command our attention develop when these hazards are magnified and adverse consequences result. If the 80 degrees is instead 102 (and especially if the humidity is high), if the thundershowers produce 3-4" of rain in just a few hours, if wind gusts are reaching 60 mph or higher, if lightning is occurring, well then we have quite a different risk profile from those in the original forecast.

Here are some examples of meteorological risks that are associated with meteorological hazards at any location:

Hazard	Weather Risk
Temperature	Heat Wave Cold Wave Wind Chill
Water	Floods Storm Surge Icy Surfaces Ice Storm Hail Drought
Wind	Tornado Hurricane Derecho High Wind Thunderstorm

Scenario Development Examples

- As an example of how broadly the scenario-building process can be utilized, here are a few, just for fun, "what if's" associated with the ongoing publication of, of all things, the *Washington Post* newspaper.

1. Early morning newspaper carriers become disgruntles, demand higher pay and later starting times, and stop delivering the paper.

2. It is found that a Pulitzer Prize award-winning series in the *Post* is riddled with everything from plagiarism to misspellings.

3. Recycled newsprint used to make the paper on which the newspaper is published cannot reach the *Post* after a ship crashes into the Woodrow Wilson Bridge.

4. Hard copy subscriptions continue their slow and seemingly unstoppable decline, leading to concerns about long-term viability of the paper.

In reality, scenario building is a serious and very important activity that many companies and organizations use to identify and address threats and hazards. Just as an oil tanker is beginning its final docking procedures, storm force gale winds begin to push the ship toward the dock.

- An alarm goes off in the alkylation unit of an oil refinery, signifying that there may be a leak of the highly toxic hydrofluoric acid.
- Water is making its way into the basement from a leak where the septic pipe exits through the basement wall.

Probability is a measure of the frequency with which a scenario is likely to occur. Probability is characterized differently depending on the amount and type of data that exists for any particular event/scenario When sufficient data of verifiable quality are available (such as for automobile accidents), probability should be characterized numerically. When there have been scattered incidents but only limited data (such as fires in nuclear power facilities), probability should be characterized in non-quantitative terms, such as "unlikely to occur" or "occurs multiple times in the course of a year". Scenario building can also identify where data is not currently available and what needs to be done to secure its availability. Additional data can improve the quality of future assessments

There are many ways to characterize the consequences of a severe weather event. And it's a lot easier to define a consequence than it is to define a probability. To provide a degree of consistency in so doing, a table has been developed as a "go-by". The table below describes the range of consequences associated with severe weather events. *This is only an example; other consequences are equally useful.* Among the criteria for defining consequence on the table are the scope of the consequences, the extent of damage to buildings and infrastructure, travel impacts, school and government closings, and injuries and deaths.

Category	Means
Building Damage	• *Damage to buildings, homes, commercial property*
Scope	• *Far Reaching* – Affecting everybody in the region • *Widespread* – affecting most in the region • *Limited* – affecting a defined segment of the region • *Localized* – affecting only a small percentage of the region
Damage	• *Major* – widespread structural damage including loss of siding, roofs and windows; trees downed on buildings, roads and telephone poles • *Significant* – localized structural damage to homes and other buildings; some trees down; but impacts either not as great as and/or not affecting as large an area as for Major • *Minor* – minor damage to homes, primarily of a nuisance variety like downed trees across a driveway or some shingles missing from roofs
	• *Light* – barely noticeable impacts
Infrastructure	• *Electrical power, telecommunications, cell service, water*
Damage	• *Crippled* – Power out for more than 3 days; mobile phones slowly shut down and are rendered inoperable; water system compromised, tap water requires boiling • *Affected* – Power out for 1-2 days; some disruption in iPhone and other telecommunication services, though malls come to life quickly • *Recovers Quickly* – temporary disruption measured in hours; most electrical outages and telecommunications issues remedied quickly • *Largely Unaffected* – few issues and of a limited basis only
Closures	• *Federal Government, Public and Private Schools, Government contractors, commerce and industry*
Extent/Length	• *Long Term* – 4 days or more; *Short Term* 1-3 days; *Few* – Some school districts; *Open* – only an occasional closure
Travel	• *Car, Metro, Rail, Air, Boat*
	• *Limited* – Roads closed or impassible; airports closed; most flights cancelled; rail at a standstill • *Difficult* – Significant delays at airports and via rail/Metro; roads blocked by snow, ice or debris and barely passable • *Minor Impact* – some delays on specific roads; minor delays at airports, via Metro and on rail system • *Back Quickly* – any travel impacts resolved quickly and with only minor delays
Fatalities/Injuries	• *Multiple*; *Some* injuries; *Few* Injuries; *None* (self-explanatory)

<u>The risk matrix is a simple, effective mechanism for visualizing risks.</u> An especially valuable feature of the risk matrix is its ability to display multiple risks simultaneously. A commonly used risk matrix has five or six levels of probability on the x-axis and four levels of consequence on the y-axis. The most probable and highest consequence risks are to the left and up. The highest risks on the matrix result both from events that have a high probability of occurring but whose consequences are relatively low to and from events where the probability of the event occurring is low, but the consequences are high. The blue, white and yellow "risk fields" represent areas broadly considered to be of equivalent risk. The matrix is primarily a qualitative tool—detailed statistics and risk determinations are not a part of this risk assessment process. Once individual risks have been placed on the Matrix, risks can be compared to one another. This allows the team to develop a seriatim of weather risks (or groups of risks) from high to low. This information is important for the next step in the process—risk management. The risk matrix is location specific. Although considered high risk, in fact major snowfalls of say a foot or more occur very infrequently.

The Climate of Northern Virginia

Northern Virginia's climate is temperate, with four distinct seasons, plentiful rainfall and some snow (sometimes just a bit) every winter. Northern Virginia's climate is also generally free of severe weather, with the most severe events (hurricanes, derechos, ice storms, major snowstorms) occurring only sporadically. For instance the risk from tornados is much higher in the Plains States, from hurricanes along the Gulf Coast and for crippling blizzards in the Northern Plains.

Dulles Airport receives about 20" of snow annually. Most of that falls during "nuisance" events of <3" of snow. There are a few instances with single snowfall totals of more than 9". In the last 20 years there have been six snow events with nine inches of snow or greater, about one every three to four years. Though few and far between, those exceptional snowfalls have been significant. At Oak Hill, VA (10 miles SE of Dulles Airport), 29.8" in 2016; 24.3" in 2010. Heavy icing, enough to bring down trees and powerlines, tends to occur very infrequently, and much less than the locals claim.

Much of the data needed to complete the meteorological Risk Matrix for Northern Virginia was readily available from the local weather forecasting office. The more data the better, accurate data of course preferred

- Number of snow events of various depths
- Number of snowstorms and ice storms per year
- Number and severity of tornados
- Number of, and total rainfall from, heavy rain events
- Number of landfalling hurricane and derecho events that have impacted the D.C. Metro areas
- Highest peak wind gust and monthly wind speed
- Record highs, lows, and high lows
- Monthly rainfall past 12 months

So, the team put together to assess the risks posed by weather in Northern Virginia will build scenarios around each specific weather event (ice storm, heavy rain or landfalling hurricane, whatever) and identify for each, using data and team member memories, the probability that such an event will occur and the consequences if it does. Each scenario is placed on the matrix, where it allows risks to be compared to each other, It also provides the opportunity for a reasonableness check of the risk assessment process just completed before transitioning to risk management.

One of the <u>highest weather risks in Northern Virginia results from the collective threats associated with thunderstorms</u>, which are numerous, and often accompanied by lightning, high winds, large hail, heavy rainfall and at times tornadic activity. Although considered to lead to only Level II consequences on an individual basis, the collective threat associated with dozens of thunderstorms each year can be substantial. It is not unusual for the high winds and lightning associated with thunderstorms to damage homes and businesses, to cause power outages and to lead to injuries and fatalities. And during any typical late Spring through September, dozens of strong thunderstorms rumble across Northern Virginia.

Another scenario is associated with <u>ice-covered steps, sidewalks and parking lots</u>, which pose a major challenge to pedestrians multiple times every winter on the Northern Tier of the U.S. Many pedestrians suffer soft tissue injuries and broken bones due to trips, slips and falls. Many more injuries in fact to pedestrians than to those in cars during the same icing events. Such icing events invariably lead to school closings or delayed openings, and traffic delays.

One of the surest ways to minimize risk from severe weather is to stay informed and to be ready to move at a moment's notice, whether it be a hundred feet to your backyard tornado shelter or fifty miles to escape. The heavy-hitters (tornadoes, land-falling hurricanes, derechos and major snow events) are capable of creating significant, crippling damage to infrastructure and threats to the public. Although the consequences of such severe weather events are significant, the frequency with which they occur in Northern Virginia is low. On the other hand, in other parts of the country where these major events occur more often (such as blizzards in the Plains States), the risk profile of the matrix there would tend to show a higher overall risk for such big severe weather makers.

So, the risk assessment process helps determine the likelihood of credible events based on data and experience. It provides structure and rigor that helps separate actual risk from perceived risk. The most effective risk assessment and management processes rely on a team approach to ensure that varied and numerous perspectives are applied to broaden the assessment process. The team should include individuals intimately familiar with the subject matter, how things are supposed to work; what could go wrong; what might happen if it does go wrong and so forth.

Consequence	Frequency of Occurrence					
	>5 per Year	1-5 per Year	1 per Year	1 every 2-4 Years	1 Every 5-10 Years	1 every 10 yrs. Or more
Major Damage, Infrastructure Crippled Gov/Schools Closed 4 Days+ Travel Limited Some Fatalities, injuries **LEVEL 1**					Derecho	Major Ice Storm Landfalling Hurricane
Damage Significant, Limited Infrastructure Affected Gov/Schools Closed 2-4 Days Travel Difficult Some Injuries **LEVEL II**	Severs Thunderstorms Icy Road, Sidewalk and Parking Lot Surfaces	F-1 to F-3 Tornado				Snow Event >15" Heat Wave
Damage Minor, Localized Infrastructure Recover Quick Gov/Schools Local Closure Travel Minor Impact Few Injuries **LEVEL III**	Rain Event >3"		Ice 0.1"-0.5" Rain vent 2"-3"	Snow Event 9"-15"		
Damage Light, Localized Infrastructure Largely Intact Government/Schools Open Travel Back to Normal Quickly Injuries Minimal **LEVEL IV**	Snow Event 1 – 3"	Snow Event 3"-9"				Drought

Risk Management reduces the level of risk defined during risk assessment by implementing measures that reduce the probability that the scenario will occur, and/or reduce the consequences should it occur. The challenge of risk management is to <u>reduce risk</u> to an acceptable level <u>without major capital, operational or maintenance investments,</u> significant change in facilities or work processes, or the need for additional human resources.

Although many risks can be reduced to near zero, investments required to reach "zero risk" typically cannot be justified. It is not possible to eliminate all risks. We live in a risk averse culture in which we often are reluctant to take any risks at all, even small ones. We let perceived risks outweigh actual risks, even when we understand the difference. Minimizing one risk may lead to the creation of others. For instance, adding a sprinkler system to your house may reduce your losses from a fire, but it also adds another threat should the water system be accidentally triggered in the absence of a fire.

Assessing meteorological risks is particularly interesting because, strictly speaking, the probability of an incipient event occurring cannot be reduced. Nothing can be done to stop a tornado or a blizzard. the more risk involves people, and especially children, the less risk we are willing to tolerate. This is true even when those perceptions of risk border on the irrational. But such factors cannot be ignored. To be effective in real-world situations, <u>risk management must incorporate solutions that are not only grounded technically, but which are socially acceptable.</u> The public needs to accept that no matter how many resources are deployed to address severe weather threats. Some risks will always be present. It is each person's individual responsibility to minimize possible impacts from severe weather threats on themselves and their families.

Addressing Risks Identified During Risk Assessment

Once risks have been identified and placed on the risk matrix, a team of experts should be assembled to identify opportunities for reducing the probability and consequence associated with the risk. It goes without saying that <u>the "solution" that provides the greatest risk reduction at the lowest cost should be considered the highest priority.</u> The table below provides some examples of changes that have been made to minimize the probability and/or consequence of severe weather events. Most of the table is accurate and many of the actions are in fact currently in place and working successfully.

Severe Event	Reduce Probability	Reduce Consequences	Status
Icy Roads	Pre-treat road surfaces with a combination of sand, salt & brine several days before icing is expected	Pre-treat road surfaces with a combination of sand, salt & brine several days before icing is expected	Widely deployed across VA . Easy to treat roads 1-2 days prior to icing event. Very cost effective & successful
	Embed temperature sensors into road surfaces to alert VDOT of impending freezing conditions	Temperature sensors indicate if snow or ice are sticking to road surfaces, or if road surface temperatures are above freezing	77 sensors currently operating on Virginia roads
Icy steps, sidewalks and parking lots	Remind public of upcoming risks for slips, trips and falls 7	Remind public of upcoming risks for slips, trips and falls	Precautionary note added to NOAA's Mid-Atlantic Winter Weather Advisories template to warn/remind pedestrians (not just drivers) of risks
Black ice	NOAA issues special weather statement to warn motorists of impending ice on roads when conditions warrant (usually just before dawn)		Implemented
High winds or ice buildup bring down power lines	Continue tree trimming year round in Northern Virginia	Continue tree trimming year round in Northern Virginia	Fully implemented; ongoing. Difficult to measure but assuredly success
		Set up mutual, reciprocal aid arrangement with out-of-state energy providers	In place
	Bury electrical wiring	Bury electrical wiring	Standard procedure for new electrical services; prohibitively expensive to retrofit
Derecho		NOAA working to improve processes for warning the public	Ongoing
		Have a severe weather response plan in place; where to go when	Training and awareness a never-ending process Has saved countless lives over the past decades

Major Snow Event		Use a work process that clears snow from roads based on an established prioritization process	Mature system in Virginia
Hurricane	Establish dedicated evacuation routes to get the public out of harm's way	Establish dedicated evacuation routes to get the public out of harm's way	For example all lanes on I-64 from the Tidewater northbound can be made unidirection8al. Gates have been installed to avoid counter current traffic
		Require stilted construction of buildings susceptible to storm surge	All new construction on, for instance, Long Beach Island, NJ required to comply
		Require building codes in vulnerable locations to have more robust roofing, siding and windows requirements	Florida among other States, adopting this requirement

So, NOAA, electrical utility companies, local jurisdictions and others have implemented a number of work practices (some obvious, some established, some new, some creative) to reduce the risks associated with severe weather. Improved building codes, year-round tree trimming, pretreating road surfaces, stilts to position homes above storm surges, dedicated evacuation routes, road surface thermometers, prioritized road-critical snow plowing, improved winter weather advisories that include pedestrian risks and black ice threats, buried electrical wiring, dedicated evacuation routes using both sides of the Interstate to maximize traffic outflow, and so forth. Although not easy to measure, the implementation of these practices has most assuredly reduced deaths and injuries, and led to less damage to buildings and infrastructure from severe weather events.

Risk management examines the assessed risks and seeks a set of cost effective changes that reduce risk probability and/or consequences. We will provide some examples of changes that are making a difference.

<u>Staying Informed –The Best Way to Lower Vulnerability</u>

When it comes right down to it, it's the public's understanding and aware-
ness of impending severe weather, coupled with the authorities' ability to provide guidance and effective evacuation, or safe-as-can-be-shelter-in-place, that
ultimately will be the biggest factor in minimizing severe storm-related deaths
and injuries. From a human risk perspective this means we need to be first, informed, then following closely thereafter, warned of expected risks. Unable to
affect the direction, intensity or path of a storm, we need to take the appropriate timely action to minimize the consequences of the severe weather. Your job
(for instance for hurricanes) is simply to get out of the way. If it's a tornado,
it's finding the safest feature wherever you find yourself—a tornado shelter in
your back yard, or your house's basement, or a bathroom or close. But what
we really need to know is how close to the real action are we?

<u>Conveying Your Risks and Vulnerabilities</u>

Yes, convey the risks, but without crying "WOLF" too often. Our TV
weather personalities have increasingly become the celebrities of the airwaves.
And increasingly they are using graphics to help the public grasp the significance
and potential damage to structures and themselves. Here is a chart, one of hundreds devised by local, county, state and national forecasters, that quite clearly
shows the threats to local community from an impending severe weather event.

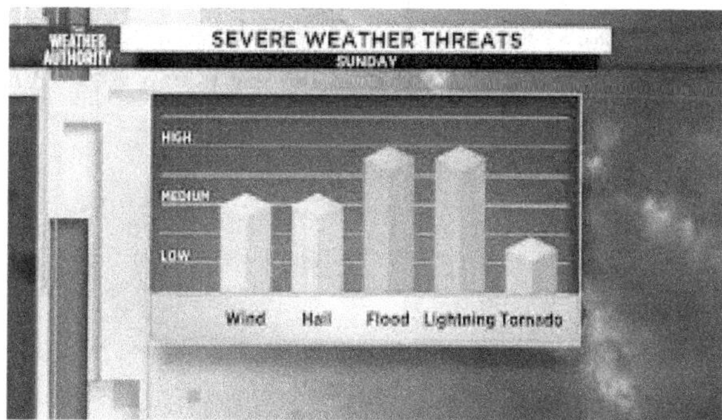

Such charts provide a simple, clear means of conveying risks to the public.

<u>What if you are asleep or the electricity is out?</u>

Many tornados occur after bedtime. No one wants to be awakened by the sound of a tornado bearing down on them. Enter a solution—NOAA Weather Radio. NOAA Weather Radio is an automated 24-hour network of VHF FM weather radio stations in the U.S. that broadcast weather information directly from a nearby National Weather Service station. NOAA began forecasting and providing warnings in the 1950s in select vulnerable cities. This was expanded to nation-wide coverage in 1967. NOAA Radio alerts are accompanied by a loud unmistakable sound that alerts the public to approaching severe weather well before the storm is close by. Your iPhone can also be programmed to receive warnings. When storms are of sufficient strength, winds can topple trees and telephone poles, leading to power outages. Television and the internet may be unavailable for many days and your iPhone will quickly lose power. Make sure you have extra batteries and/or fully charged "power bars" that will greatly extend the useful life of your phone.

<u>NOAA's Natural Hazard Risk Assessment</u>

People instinctively and invariably want to understand if, where and how they are vulnerable to severe weather events, and how those risks compare to other parts of the country. Is there really a safe place to live? Before moving to any new location, it's smart to understand the severe weather risks inherent in that area.

NOAA, in conjunction with FEMA and independent experts, has produced a very detailed study called the Natural Hazard Risk Assessment which provides a seriatim of losses to human life and property from highest to lowest in every one of the roughly 3000 U.S. counties. It is based on a number of factors that include the value of each county's tangible assets (homes, barns, livestock, infrastructure) and the concurrent susceptibility to loss, based on the historic occurrence of losses from such hazards. The work process includes a couple of unique human behavioral factors called "Social Vulnerability" and "Community Resilience." The goal of this effort was to assign a numerical "Risk Index" to each county. The "equations" that are used to develop the Risk Factor are shown in the table below.

	X	Natural Hazard Frequency	X	Historic Loss Ratio	=	Expected Annual Loss
The buildings, population, and agricultural value exposed to a natural hazard event		The estimated annual frequency for a given severe weather event to happen		The proportion of those values that have historically been impacted by natural hazards		The expected annual dollar value loss due to natural hazards
Expected Annual Loss (see above)	X	Social Vulnerability	db*	Community Resilience	=	National Risk Index
The expected annual dollar value loss due to natural hazards		The susceptibility of adverse impacts from natural hazards on the general public		The ability of a community to prepare for, absorb, recover from and adapt to, the impacts of natural hazards		Potential for negative impacts resulting from natural hazards

*db = divided by

Here are some of some examples of U.S. county risks and precursors derived directly from the NOAA Risk Index publication.

County	Population	Value	Expected Annual Loss	Societal Vulnerability	Community Resilience	Risk Index
San Bernardino CA	2.04M	180T	Very High	Moderate	Moderate	Very High
Fairfax VA	1.4M	157T	Moderate	Very Low	Rel Low	Very Low
Kent TX	808k	139M	Very Low	Rel High	Rel Med	Very Low

The Natural Hazard Risk Assessment process was prepared to help planners, emergency managers better understand what hazards are most likely to occur in their counties. The Index can help enhance risk mitigation plans, identify the need for more thorough risk assessments, and prioritize and allocate resources The concept is that the better informed the public and emergency response personnel are, the faster and more effective their responses will be, and the more protection that can be applied to reduce losses to human lives and infrastructure.

For the purposes of this study NOAA identified these natural risks:

Avalanche, coastal flooding, cold wave, drought, earthquake, hail, heat wave, hurricane, ice storm, landslide, light river flooding, strong wind, tornado, tsunami, wildfire and winter weather

Ultimately an overall Risk Index Score and individual hazard score are calculated for each county.

Managing Risk – Hurricane Vulnerabilities

Let's briefly look at a few additional resources that will help us understand and react to warnings of severe weather. Climatological data has been collected on hurricanes since about 1850. Since that time just short of 300 hurricanes have made landfall in the United States from Eastport, Maine to Brownsville, Texas. Surprisingly only about two landfalls per year. By noting where each hurricane has made landfall and its strength upon coming ashore, it is possible to construct 10-, 30- and 50-year probabilities for Category 3 or greater U.S. hurricane landfalls.

This data shows that, for instance, Dare County, NC, south of Cape Hatteras, has a 91% probability of being struck by a Category 3+ hurricane over a 50-year period. The risk presented by such a threat should lead to a number of risk management strategies to reduce the consequences of such an occurrence. This would include the implementation of beefed-up building code requirements including overall design, roofing materials, window shuttering, stilts and strengthened internal structural features. Or it could lead to what might be the best solution of all—a prohibition of building anything at all on flood plains and barrier islands. Similar risks can be calculated for other meteorological phenomena such as blizzards and derechos. Sometimes it makes sense to take a step back and get away from the detailed risk assessment and management tools—the excruciatingly detailed and possibly confusing NOAA Natural Hazard Risk Index with its hard to explain social vulnerability and community resilience factors. And dispense with the risk matrix for a second. A simple accumulation of severe weather data in various parts of the country can be very instructive. Let's assume you were looking for a relatively low cost of living place in which to retire. And let's assume some searching found that central Oklahoma offers very affordable housing and a climate featuring only a few very cold days annually. But a closer look tells you that towns like Tulsa and Oklahoma City are also among the top cities in the country likely to experience tornados and hail, year in and year out.

Top Three U.S. Cities With	
Propensity for Tornados	Oklahoma City, OK
	Tulsa, OK
	Dallas, TX
Highest Annual Snowfall	Marquette, MI
	Sault Ste Marie, MI
	Syracuse, NY
Coldest Temperatures	International Falls, MN
	Duluth, MN
	Caribou, ME
Highest Annual Rainfall	Hilo, HA
	Quillayute, WA
	Astoria, OR
Propensity for Hail	Tulsa, OK
	Amarillo, TX
	Oklahoma City, OK
Propensity for Tropical Weather	Cape Hatteras, NC
	Communities on Florida's Atlantic Coast

WEATHER POSTSCRIPTS

I hope you enjoyed reading *Weather Headliners* as much as I enjoyed writing it.

I would like to share some special weather-related information gleaned from over 50 years of observing, learning, teaching and otherwise sharing weather information. First, there is a collection of characteristics of weather which I have gathered from ordinary people all around the world. You will see characteristics like beautiful, awe-inspiring, quiet, fascinating and complicated.

Second is an article called "What I Know". I have learned a lot about the weather by just observing it. The sky and the clouds, wind speed and direction, when storms start and end, the size of snowflakes. "What I Know" about the weather has been fine-tuned from hundreds of observations, especially in southern New England and the mid-Atlantic States, where I have lived for many years. I have learned much about weather observation, forecasting, models and balloons. Whatever you decide to do, don't forget to go outside a few times a day. You might be surprised to know how much valuable information can be gleaned from the sky and the clouds it contains.

December 4, 2023

CHARACTERISTICS OF THE WEATHER

No matter how much you do or don't follow the weather, have a favorite app on your iPhone that "instantly" makes you aware of current weather conditions and impending severe weather, or could care a hoot about worldwide global climate change, the weather is a fascinating subject. Here are some characteristics of weather I have made note of. Retirement gave me the time to write them all down. It also gave me the time to write *Weather Headliners*.

Here is "What I Know":

- Weather is always present.
    - o You can't escape it.
    - o No matter where you are (inside or out) the weather is right there with you.
    - o When you take a trip, the weather is waiting for you when you arrive.
- Weather is on everyone's mind.
    - o More times than not, and especially if something meteorologically significant has occurred, is occurring now, or is expected to occur, weather is the first topic people will mention in casual conversation.
    - o And depending on how significant the event is, they will very likely continue to talk about it for as long as you will listen.
- Weather is fascinating.
    - o Cloud bursts.
    - o Lightning illuminating your bedroom in the darkness of the early morning hours.
    - o Thunder slowly rumbling away, or rattling the house with ear-splitting intensity.
- The weather is ever-changing.
    - o A measurable snowfall less than 24 hours after temperatures were in the 50s.

o Long entrenched drought suddenly eased by rain from the remnants of a tropical cyclone.

- Weather is beautiful.
    o Snowflakes gently drifting to Earth, or swirling in the nighttime wind when the backyard light is turned on.
    o Ice-covered trees sparkling and glistening in the morning sunlight
    o Rapidly advancing cirrus clouds, slowly but surely filling in the sky as a snowstorm approaches
    o Angry black clouds swirling directly above your deck as you watch a severe thunderstorm approach
    o A rainbow in the eastern sky as a storm recedes
    o The sun setting in a blaze of oranges, yellows and reds
- Weather is anticipation.
    o That special feeling—a combination of awe, uncertainty, fear and concern—that precedes the approach of a hurricane, a severe thunderstorm, a snowstorm, an icing event
    o The first meaningful cold front, arriving to push away the heat and humidity of the summer
    o Fall foliage, the first frost, an early November snowfall, a white Christmas
- Weather is quiet.
    o A steady snowfall on a windless evening, deadening the noise of a busy neighborhood.
    o Still, quiet, serene; the world seemingly at peace
- Weather is simple.
    o Just give me the bottom line—temperature? Now and later today? Is it going to rain? When? Light? Heavy? Should I worry about the wind?
- Weather is complicated.
    o El Nino, occluded fronts, dry slots, the Pacific Decadel Oscillation, ACE, hailstone formation, the Collision-Coalescence theory of raindrop formation
- Weather is capable of eliciting the full range of human emotion.

o Awe, respect, wonderment, concern, fear, intrigue, uncertainty, beauty, fascination, reverence
o And sometimes the best weather is no weather at all. You know, those evenings when you sit outside, perhaps at a restaurant, and the temperature is so perfect, the humidity non-existent, the wind so light, that you really have no sensation that there actually is any weather.

A Young Boy's Fascination of the Weather

It seems that more and more people are developing an interest in some phase of meteorology. Whether it's tracking a Category 3 hurricane that just entered the Gulf of Mexico, looking at all the video from the recent catastrophic flooding in Vermont, going on line to look at increasingly understandable numerical weather prediction (NWP) techniques, or just watching the Weather Channel, people are getting involved with weather. This phenomena spans generations and ages. Following is an essay called A Young Boy's Fascination with the Weather

I have been fascinated by the weather for as long as I can remember, maybe no time more pronounced than aa a youngster growing up in Milford, Connecticut. These are some of my most special experiences and memories

- Lying on the cool green grass of our little front yard, watching the clouds ebb and flow, shrink and grow
  - Or anticipating the changing of the seasons, excited about the variety of weather invariably accompanying each
- Or waiting for a heavy snowfall, maybe heavy enough for a snow day, no school; glued to the radio for the latest forecast; maybe they'll close Bryan Hill Road so sledders can glide all the way to the bottom, where the police have placed those wooden A-frame barriers to remind the sledders, their minds lost temporarily in the exhilaration of the descent, that there'll be traffic on Buckingham Avenue
- Or realizing it's not going to snow much, if at all, listening to the rat-a-tat-tat sound of sleet pellets, signaling the end of the snow event, a measly less than an inch, surely not enough to call off school. Then there are the lucky ones just 20-30 miles north of here in Meriden or Hartford who might awake to 6 to 8 to 12 inches

- Or freezing our butts off at the annual fourth of Juily fireworks celebration at Pond Point beach, unprepared for the cold onshore winds that even a well-wrapped native couldn't understand

- Or skating on our own little, shallow frozen "pond" at the bottom of Olive Street. On very cold days putting on and lacing up my skates in the warmth of my house and literally skating down to the pond. The pond was our own little creation and we lovingly shored up its walls each October, waiting for Fall's rains to fill it up. And, if we were lucky, if the rain had been plentiful and the winter cold, we could wind our way between the trees on a spectacular skating trip all the way to Melba Street

- Or experiencing the decade of the 1950's when Long Island Sound caught more than its fair share of hurricanes. We would run along the beach, parallel to the shoreline, just as the water from the Sound began to overflow the beach at Bayview, running between and standing on those old concrete chairs, using them as safe havens as the water, becoming increasingly deep, rushed below us

- Or wondering if this storm would cause low-lying streets in the vicinity to flood again? Huge waves, at least by Long Island Sound standards, hitting the concrete retention walls that separated summer homes from storms' fury. Watched the spray shooting up what seemed like 30-40 feet, preceded by the strangest "puoff" sound as water slammed against the embankment and spray shot high into the air

- Or sitting in the living room, waiting for a thunderstorm to materialize, hearing it so faintly at first, so far away, using the one-thousand-one, one-thousand-two rule, counting the seconds between the flashes of lightning and rumbles of thunder, judging how far away it was; was it coming closer or sliding to our north, away from us?

- Or lying on the bed at Aunt Ruth's house, hearing the sounds of branches scraping the siding and that same wind rustling through the tree leaves, creating wonderful sights and sounds, some a little scary

- Or visiting Ansonia in 1955 after hurricanes Connie and Diane, making U.S. landfalls just a week apart, had unleashed so much rain that every bridge over the Naugatuck and Housatonic rivers from Seymour to the Sound was damaged or destroyed

- Or, from a perch well above the town, finding Ansonia filled with water up to the second story windows

- Or listening with rapt attention and awe as my parents recounted their encounters with the great New England Hurricane of 1938, before they were married, stories that included my father, in New Haven at the time, experiencing the fury of the storm as the eye passed nearby and my mother, eventually reaching the family's beach house in Milford, only to find the storm had literally lifted it off its foundation and deposited it on the beach

- Or building in the basement of our little Cape Cod, a huge paster of paris depiction of the water cycle for my eighth grade science project, making sure it wasn't too big to fit through the door, or too heavy to carry out

- Or, finally, building my own rudimentary but "functional' weather station out of predominantly common household items, allowing me to measure wind speed and direction, rainfall, temperature and humidity, just like real meteorologists

WEATHER HEADLINERS—WHAT I KNOW

Here then are some of the things I just know from my observations of, and learning about, the weather.

- You can't have Indian Summer, that wonderful spell of warm weather closer to winter than summer, until you have had a frost first.
- When you see the underside of leaves on a hot, humid summer afternoon, the process that initiates thunderstorms is underway.
- Forecasters should take at least one look at the sky before finalizing their forecasts.
- The so-called backlash, an additional band of heavy snow that comes around a nor'easter after the primary storm has passed you by, is frequently forecast, but rarely materializes.
- The appearance of large snowflakes gently falling to earth signals the end of the snowstorm.
- For many who live just a few miles from the Atlantic Ocean, storms from the south frequently start out as snow then quickly transition to sleet then plain rain as warmer air filters into the storm.
- The blacker the cloud, the darker the atmosphere, the higher the cloud tops, the heavier the rain, the more severe the lightning, and the more likely hail will occur.
- During the winter, the temperature of the ground, and consequently that of road surfaces, are very important considerations for drivers.
- When talk turns to how hot it was last summer, or the amount of snow that fell last winter, or how rainy it has been this month, many people will tell you with certainty that they remember the details. They are usually wrong.
- Rain is the parameter that people have the most difficulty estimating.

- Snow is the parameter that people most love to exaggerate.
- There is one right way and lots of wrong ways to measure snow depth.
- It can snow when the air temperature is 40 degrees, and rain when it is 27.
- Drought begets drought.
- The deepest snowfalls start early and heavy, working off high energy fields well ahead of the actual low pressure system
- A corollary reads like this—if heavy snow is predicted and the initial snowfall from the low pressure system is light and sporadic for hours, the snow-making energy never took hold, or passed you by.
- A brightening sky, usually to the south and west, invariably heralds the end of a major snow event.
- If it is raining, or has rained, and it is still dry under a tree, it hasn't rained a tenth of an inch.
- I don't subscribe to the accuracy of all the weather sayings, but "ring around the sun or moon means rain or snow upon you soon" seems to work.
- The biggest mistake one can make is to forecast the future state of the weather the next hour, the coming afternoon, next weekend, next winter, based on the current weather conditions.
- More is left to be understood about the weather than is understood today.
- I suspect a lot of you share my interest in, and love for, the weather.
- If you weren't one of those before who read weather headliners, I hope you are now

Weather Headliners – Reference List

G Falkovich, G, Belkovsky B, Foxon A; Phys Rev.Lett.86.2709 (2010)

Linacre and Geerts

Wizemann Institute of Science

Omanovic, Najda et al

University of Vermont Libraries, Center for Digital Initiatives.

Fox Weather

Halverson, Jeffrey S. University of Maryland

Strader, Stephen M & Walker, Ashley S, Northern Illinois University

Lamb 1969/Schneider & Mars 1975

McIntyre, Michel Edgeworth

Washington Post

www.ingramcontent.com/pod-product-compliance
Lightning Source LLC
Chambersburg PA
CBHW051515150726
47997CB00001B/262